Weather in the Lab
Simulate nature's phenomena

Thomas Richard Baker

TAB Books
Division of McGraw-Hill, Inc.
New York San Francisco Washington, D.C. Auckland Bogotá
Caracas Lisbon London Madrid Mexico City Milan
Montreal New Delhi San Juan Singapore
Sydney Tokyo Toronto

Allen County Public Library
900 Webster Street
PO Box 2270
Fort Wayne, IN 46801-2270

FIRST EDITION
SECOND PRINTING

© 1993 by **Thomas Richard Baker**.
TAB Books is a division of McGraw-Hill, Inc.

Printed in the United States of America. All rights reserved. The publisher takes no responsibility for the use of any of the materials or methods described in this book, nor for the products thereof.

Library of Congress Cataloging-in-Publication Data

Baker, Thomas R.
   Weather in the lab : simulate nature's phenomena / by Thomas R. Baker.
     p.   cm.
   Includes index.
   ISBN 0-8306-4309-5.   ISBN 0-8306-4307-9 (pbk.)
   1. Weather—Experiments. 2. Weather forecasting—Experiments.
I. Title.
QC981.B315 1993
551.5'078—dc20   92-34050
                                                  CIP

Acquisitions editor: Kimberly Tabor
Editorial team: Sally Anne Glover, Editor
               Steve Bolt, Executive Editor
Production team: Katherine G. Brown, Director of Production
                 Rose McFarland, Layout
                 Susan E. Hansford, Typesetting
                 Kelly S. Kristman, Proofreading
Design team: Jaclyn J. Boone, Designer
             Brian Allison, Associate Designer
Cover design: Holberg Design, York, Pa.   TAB 1
Cover illustration: Steve Parke, Baltimore, Md.   4332

Dedication

To my beautiful seven-year-old daughter, Noel, who is always asking, "Dad, what cloud is that?" Also for the nighttime stories we tell each other, where "the sun is always shining on Ponyland."

To Dr. George Fischbeck of KABC Television in Southern California. His support of me in Operation Desert Storm and his encouragement to pursue the writing of this book have been tremendously appreciated.

Also to Mr. Henry Rollins, science teacher at Westlake High School in Thousand Oaks, California. His quiet, supportive assistance and contribution as a co-presenter for our seminars at the National Science Teachers Association conventions gave the spark to all of this.

Contents

	Preface	ix
	Introduction	xi

THE SUN

1	Diameter of the sun	2
2	Isotherms in the classroom	7
3	Radiation, convection, & conduction	12
4	Heat in the atmosphere	18

THE EARTH

5	Temperature & evaporation of water	24
6	Heating by convection	27
7	Determining dew point	31

THE ATMOSPHERE

8	The barometer	38
9	Calculating the size of a raindrop	42
10	The relative humidity indicator	46
11	The wind chill factor	50
12	The weight of the atmosphere	56
13	A convection cycle	58
14	The cold front	61
15	The occluded front	65
16	The temperature inversion	68
17	Measuring the oxygen content of air	71
18	Air exerts pressure	74

GEOPHYSICAL FEATURES

19	Heating & cooling of model land forms	78
20	Heating & cooling of actual land forms	84
21	How mountain ranges affect climate	89

WEATHER FORECASTING

22	Long-term weather observations and graph	94
23	Satellite pictures from the newspaper	99

24	"Spot" 24-hour weather predictions	102
25	The weather data station	105

APPENDICES

A	Tracking storms of the season	107
B	Weather trends	111
C	Environmentally speaking	123
D	Required equipment	128
E	Manufacturers of laboratory equipment	130
F	Answers to lab questions	132
	Index	144

Preface

Weather is certainly all around us. It begs for our attention in all of our activities. We purchase clothing, air conditioners, fireplaces, and air filters to help keep the weather in check. Yet, we still can't forecast the weather with 100 percent accuracy.

It will be education and experimentation at all grade levels (as well as continued exploration in industry) that will advance the quality of the study and forecasting of weather. I wrote this book as part of this quest for quality in the laboratory. Designed for middle-school and high-school students of Earth Science and Physical Science, this book will "handshake" logically with many texts on the earth sciences; each lab models one or two key features of an outdoor "real-world" weather phenomenon. In one class period, the student will set up and run an experiment, acquire data, and interpret this data to further explain and understand a facet of weather.

Industry and business supervisors today want employees to work well with one another. This book attempts to address that desire. Most of these exercises place students into groups of two to four, and the group works together somewhat autonomously to engender the crucial skill of "getting along."

This book also helps in the study of mathematics—specifically, algebra and geometry. Graphing and analyzing data is heavily stressed in specific lab exercises.

Each lab has questions and suggestions for further pursuits. Because weather is not exactly the same everywhere, each student will investigate slightly different occurrences. These nuances over a large geographical area give the study of weather a special "flavor" not readily enjoyed by other sciences.

Since science is not exact, this book is merely an avenue for further exploration. I hope that other teachers and their students will develop their own investigations to enhance the continuing changes that weather and its study are undergoing.

Perhaps you won't reach that enviable goal of forecasting with perfect accuracy. However, when a model of the atmosphere exists that's receiving constant data from a global observation system, you'll be that much closer to truly understanding the behavior of our earth's atmosphere.

Introduction

Our weather is composed of four influencing factors: the sun, the earth, the atmosphere, and geophysical features. Each of these facets is an actor playing a most convincing part. From the searing hydrogen fusion furnace on the sun to the rippling, wind-influencing features of a mountain range on earth, these four highly interrelated companions drive the "engine" of climate, including cold, heat, rain, snow, and violent storms.

Remove any one of these four "engine parts," and the machinery of our weather would soon shut down. With no sun, the earth would be encased in ice in a short period of time. Without the earth, obviously there would be no life; the absence of an atmosphere on earth would guarantee nonexistence of all life forms, and an earth without geophysical features would have featureless weather—it would be the same across the entire planet.

The weather-related labs in this book are categorized according to which weather factor they mostly associate with—the sun, the earth, the atmosphere, or geophysical features. The labs are designed for grade levels eight through twelve, and students can do these exercises to bring out the full "flavor" of the Earth Science text they're currently using. Because some teachers might not have all the equipment for both methods, two different procedures are provided for Lab 1. The two procedures show how different models can produce similar results.

Also, to put it all together, the last part of the book deals with weather forecasting, perhaps the most talked-about aspect of weather, and the most enjoyable. Will you think these labs are fun? Most certainly—for the science of meteorology is the most accessible and challenging of all the sciences because it affects every one of us.

LAB SAFETY PROCEDURES

1. Wear goggles or safety glasses when any type of heating occurs.
2. When preparing glass tubing or thermometers for insertion over rubber stoppers, use glycerin or liquid soap and water.
3. Avoid looking at the sun or its reflection from glossy or shiny surfaces.

4. Observe all manufacturer safety precautions in the handling of all chemicals—liquid or solid. Wash hands thoroughly after leaving the laboratory.
5. Observe all safety precautions with heating devices/lamps, and handle all heated materials with care.
6. Be sure electrical cords are coiled away from any sink or pans filled with water.
7. Use masking tape strips to tape thermometers to the glass walls of an aquarium.
8. Wear goggles when making a smoke source (such as burning newspaper).
9. Wear gloves or use towels when handling any glass panes.
10. Clean all spills of water with towels or a mop.
11. Dispose of broken glassware in appropriate containers.
12. Dispose of solid waste in appropriate containers, not in the sink.
13. During periods of violent weather, all weather forecasts and observations must be made with extreme caution and personal safety in mind.
14. Report all accidents, spills, and injuries in the laboratory (no matter how small) to your instructor at once. Avoid contact with chemicals.
15. Completely follow all directions given by your instructor.

ABOUT THE SAFETY ICONS

Throughout this book, you'll find special safety icons. These icons mean the following:

 Students need adult supervision for this activity.

 Be careful of the heat and/or flame.

 Be careful to avoid burns and/or cuts.

 For safety, you need special equipment such as gloves or goggles.

 Do not look at the sun. Eye damage can result.

GRAPHING HINTS

1. Always use a straightedge that's longer than the paper you're using.
2. Always use a number two pencil.

3. Use graph paper that's four squares to the inch.
4. Make the coordinate axes for both the X and the Y five boxes in and five boxes up on the graph sheet.
5. Choose an appropriate scale to completely fit the data you've recorded.
6. For multiple lines, use a set of colored pencils.
7. For multiple Y axes, see FIG. 22-5 on page 97 for an example.
8. Label all X and Y axes with the appropriate titles and units used.
9. Be sure to title the graph you've made, preferably in ink.
10. Don't write anything on the reverse side of the graph—this ensures readability by your lab partners and instructor.
11. The lines on a sheet of graph paper can be used for any scale. However, don't subdivide more closely than one-half of a box.

LAB SIZE & LENGTH

The following table shows the recommended size for lab groups and the recommended number of class periods for each lab.

Recommended group size and lab length

Lab	Number of students	Periods
Lab 1	2	1
Lab 2	All	1 or 2
Lab 3	Demonstration	1
Lab 4	4	1
Lab 5	4	1
Lab 6	Demonstration	1
Lab 7	2	1
Lab 8	2	1
Lab 9	2	1
Lab 10	2	1
Lab 11	2 to 4	1
Lab 12	2 or demonstration	1
Lab 13	2 or demonstration	1
Lab 14	Demonstration	1
Lab 15	Demonstration	1
Lab 16	Demonstration	1
Lab 17	2 to 4	1
Lab 18	1 to 4 or demonstration	1
Lab 19	All	2
Lab 20	2 to 4 or demonstration	2
Lab 21	2 or demonstration	1
Lab 22	All	As needed
Lab 23	All	As needed
Lab 24	All	As needed
Lab 25	All	As needed

For grade levels lower than middle school, these exercises should either be performed under strict supervision by an adult or as a demonstration by the instructor.

PART 1

The Sun

Containing 99 percent of all matter in our solar system, the sun converts four million tons of its own mass into light every second. The earth receives only two-billionths of this light. However, it's this light that sustains us and is the engine for our complex weather machine.

In five billion years, the sun will lose its ability to sustain its nuclear fusion furnace and will swell into a red giant. Stoking its "fires" with what little fuel remains, the sun will eventually shrink into cold ash and shine no more. By that time, humans probably will have left the earth for more hospitable living conditions elsewhere.

The following four labs give an approximate idea of the size and power of the sun. *Use caution with Lab 1; do not look directly at the sun or its reflection.* Labs 2, 3, and 4 will show you how to measure the sun's influence on the temperature of the air.

1
Diameter of the sun

Using already known physical constants (earth-sun distance) and the diameter of the image that you'll cast on a card or classroom wall, a mathematical proportion will give you a fairly accurate estimate of our life-giving star's diameter.

PURPOSE
The purpose of this lab is to approximate the sun's diameter.

MATERIALS NEEDED (METHOD ONE)
- Meter stick.
- Pin.
- Cardboard.
- White paper.
- Glue stick.
- Hand-held hole punch.
- Pencil.
- Scientific calculator (exponential capability).
- Scissors.

METHOD ONE PROCEDURE
1. See FIG. 1-1.
2. Using the hole punch, carefully punch a hole in the center of a 5-centimeter by-5 centimeter cardboard square.

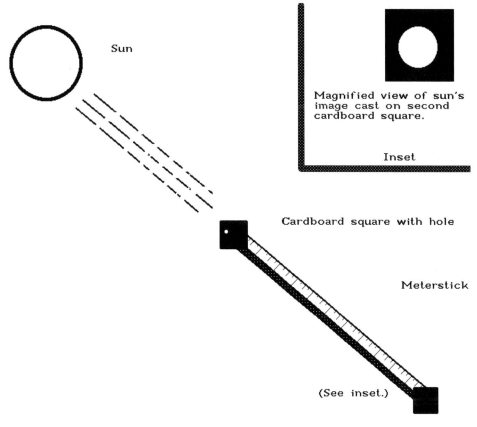

1-1 Method one setup

3. Cut a 3-centimeter-by-3-centimeter square of white paper.
4. Glue over the cardboard square.
5. Make a pinhole in the white paper square through the hole in the cardboard square.
6. Glue another square of white paper (3 centimeters by 3 centimeters) over another 5-centimeter-by-5-centimeter square piece of cardboard.
7. Tape both cardboard squares at opposite ends of the meterstick.
8. Be sure they're aligned with each other.
9. Go outside. *Do not look at the sun or its reflection! Eye damage can result!*
10. Stand up the meterstick until there's no shadow cast on ground by the stick.
11. The paper with the hole should be pointed at the sun.
12. Center the faint circle on the bottom of the cardboard square.
13. Mark the approximate diameter of the faint circle on the bottom of the card.
14. See FIG. 1-2 to calculate the diameter of the sun

DIAMETER OF SUN (km)	DISTANCE TO SUN (km)	
x =	1.495×10^8 km	
Diameter of faint circle (km)	Distance to faint circle (km)	
	Method 1	Method 2
your value =	0.001 km	km

Example proportion: $\dfrac{\text{Diameter of Sun (X)}}{\text{Distance to Sun}} = \dfrac{\text{Diameter of Circle}}{\text{Distance to Circle}}$

Cross multiply to solve for X.

1-2 Data table for methods one and two.

MATERIALS NEEDED (METHOD TWO)
- 50-meter measuring tape.
- Large, square mirror—6 inches by 6 inches (15 centimeters by 15 centimeters).
- White paper.
- Hand-held hole punch.
- Pencil.
- Scientific calculator (exponential capability).*
- Scissors.

METHOD TWO PROCEDURE

1. See FIG. 1-3.
2. Punch a ¼-inch hole into the center of a 6-inch square sheet of paper.
3. Tape this paper to a 6-inch square mirror.
4. Go outside. *Do not look at the sun or its reflection! Eye damage can result!*
5. Stand the mirror against a wall, gate or other structure so the reflection of the sun will be sent back to the inside of the room.
6. Record the distance from the mirror to the wall in the classroom. (See FIG. 1-4.) Note: The earth's rotation will move the image after several minutes. Readjust the mirror to maintain the image of the sun in the classroom.
7. See FIG. 1-2 to calculate the diameter of the sun.

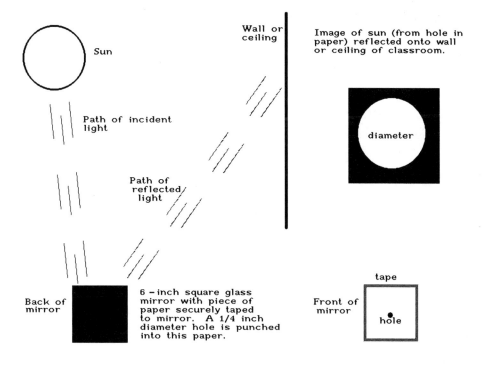

1-3 Method two setup.

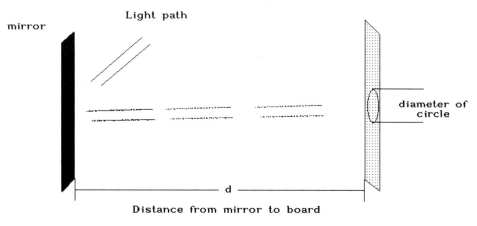

1-4 Student data

Diameter of the sun

QUESTIONS/CONCLUSIONS

1. The actual diameter of the sun is 1.393×10^6 kilometers. How does this compare to your calculated value?
2. Why is the sun a circular shape?

POST-LAB ACTIVITY—LAB I

Using a plastic or metal washer of any size, estimate the diameter of the moon. Hold the washer at arm's length. Sighting to the moon (preferably at moonrise) with one eye closed, the moon should be visible through the washer. Bring the washer slowly back to the eye. At the point where the moon's circular surface just touches the inside circumference of the washer, have a lab partner measure the distance from the washer to your eye. Record this in centimeters. Next, convert this value to kilometers. Record the diameter of the washer in centimeters and convert it to kilometers. Using the distance the moon is from the earth (3.862×10^5 kilometers) calculate the diameter of the moon. (How close is your value to the actual value of 3.4763×10^3 kilometers?)

Use a proportion calculation similar to the one you used for the sun diameter. Be sure to use ratios similar to these:

$$\frac{\text{diameter of moon (X)}}{\text{distance to moon (km)}} = \frac{\text{diameter of washer (km)}}{\text{distance from eye to washer (km)}}$$

Your answer for X (diameter of the moon) should be in kilometers (km).

2
Isotherms in the classroom

A typical weather map has lines of pressure drawn on it. Your "map" of the classroom will have lines of temperature drawn on it. In particular, your room will have its own unique microweather. Countless drafts, open windows, doors, and movement by students and air will show that the temperature of a classroom is certainly an arbitrary notion.

PURPOSE

The purpose of this lab is to observe the different layers of temperature throughout a room's area and height.

MATERIALS NEEDED

- 27 thermometers.
- 27 students.
- Pencils (red, blue, green—use a suitable number.)

PROCEDURE

1. See FIGS. 2-1, 2-2, and 2-3. Make copies as needed.
2. Arrange students to work in groups of three.
3. If there aren't enough students, select key areas in the classroom to measure.
4. Have one student in each group be at floor-level, one student be at desk-level, and one student be at above-desk-level temperature stations.
5. Leave a door or window open to provide interesting temperature effects.
6. After 10 minutes, have each student record the temperature at his or her station.

Classroom Temperature Table

STATION:	①	②	③
STATION:	④	⑤	⑥
STATION:	⑦	⑧	⑨

Floor-level temperatures
(degrees Celsius)

2-1 Floor-level student chart

Classroom Temperature Table

STATION:	⑩	⑪	⑫
STATION:	⑬	⑭	⑮
STATION:	⑯	⑰	⑱

Desk-level temperatures
(degrees Celsius)

2-2 Desk-level student chart

8   The sun

Classroom Temperature Table

```
STATION:    (19)         (20)         (21)
            ____         ____         ____


STATION:    (22)         (23)         (24)
            ____         ____         ____


STATION:    (25)         (26)         (27)
            ____         ____         ____
```

**Above-desk-level temperatures
(degrees Celsius)**

2-3 Above-desk-level student chart

7. Use red pencil for floor level, blue pencil for desk level, and green pencil for above-desk level.
8. See FIGS. 2-4, 2-5, and 2-6 for sample temperatures of similarly recorded data.
9. Graph recorded data using the following recommendations:
   ~ Use a different colored pencil for each level and connect all equal temperatures with a smooth line. This will form an *isoline* (a line connecting equal temperatures).
   ~ Each end of the isoline may connect only at the walls.
   ~ The isoline may not cross other isolines.
   ~ The isoline may not touch other isolines.
   ~ When the temperature is completely surrounded by different temperatures, draw a ring around that station's temperature to show that that's the only temperature in that region of the room.

OBSERVATIONS

1. As elevation in the room is increased, does the temperature increase or decrease?
2. How does the temperature at the floor near the door entrance compare to the back of the classroom opposite the door?

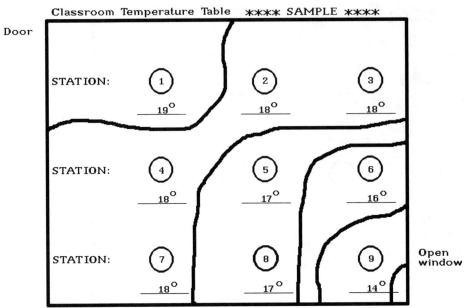

2-4 Example floor-level chart

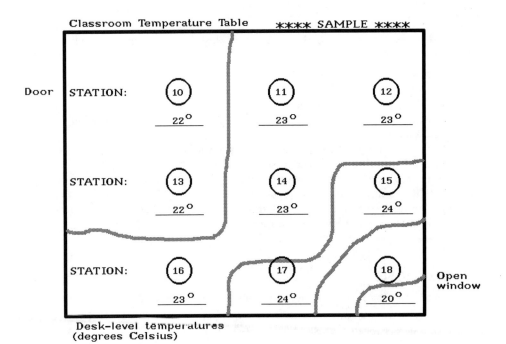

2-5 Example desk-level chart

10 The sun

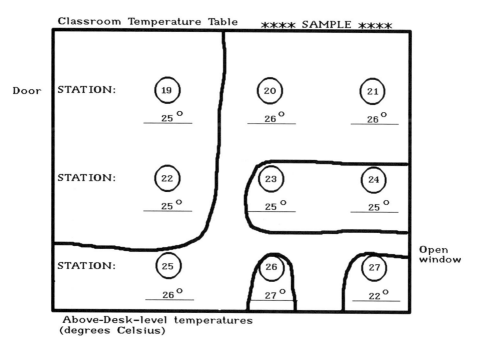

2-6 Example above-desk-level chart

QUESTIONS/CONCLUSIONS

1. Examine an isotherm map (on TV news or in a newspaper). How is your map similar to these? How is it different?
2. Where does all the data come from to produce such a large isotherm area (such as your region or the United States)?
3. Why is the air in a "hot air" balloon heated?

POST-LAB ACTIVITY—LAB 2

Create an isotherm map of a portion of your school grounds. Use a sunny or cloudy, calm day. If you have a two-story or three-story building, this will give more interesting results. Draw a scale map of several buildings and grassy areas. (Use graph paper to maintain accuracy and neatness of the drawings.) Place station numbers where students will stand and record their temperatures. Make enough copies of this map for the students. Pass out thermometers among at least 20 students. Spread the students out along the grounds. Wait ten minutes, then record temperatures.

Plot the temperatures in the same manner used for the classroom isotherms. Draw smooth lines connecting the temperatures that are equal or nearly equal. Use a different colored pen or pencil for different levels.

On some days, the outside temperature will rise as elevation is increased because of the possible existence of an inversion layer. (An inversion occurs when a layer of cooler air is "trapped" by a higher layer of warmer air.)

Also see question 59 and 60 in appendix C, "Environmentally Speaking."

3
Radiation, convection, & conduction

Radiation, convection, and conduction are responsible for the "machinery" of weather on our planet. Air, water, land, and buildings all absorb (or radiate) the sun's energy through one or more of these processes.

PURPOSE
The purpose of this lab is to define radiation, convection, and conduction and to observe an operating example of each term, using popcorn as the heated substance.

MATERIALS NEEDED
- Microwave oven.
- Packaged popcorn that can be microwaved.
- Hot-air popper.
- Loose popcorn.
- Hot plate.
- Packaged popcorn for stove-top use.

PROCEDURE
Caution: Observe all safety precautions with materials and heating devices.
1. Define radiation.
2. See FIG. 3-1 for an illustration of radiation.
3. Pop the popcorn in the microwave oven. *When opening the bag, be careful of the escaping steam.*
4. Define convection.

12 The sun

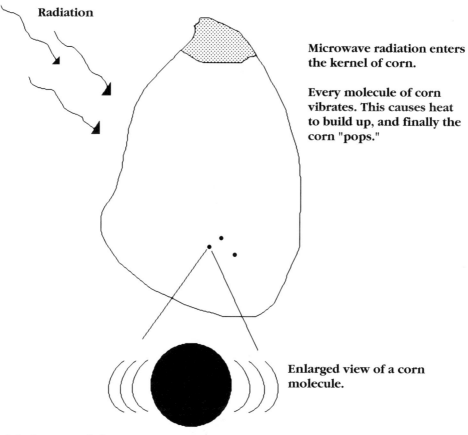

3-1 Popcorn radiation

5. See FIG. 3-2 for an illustration of convection.
6. Pop the popcorn in the hot-air popper.
7. Define conduction.
8. See FIG. 3-3 for an illustration of conduction.
9. Pop the popcorn on the hot plate. *Again, be careful of the escaping steam.*
10. Enjoy eating the results.

Figures 3-4 through 3-7 show the process of energy heating a steel bar.

OBSERVATIONS

1. Which popcorn container is the coolest? Why is this so? What was able to penetrate the paper bag in the microwave?
2. Which container was the hottest? What made it so warm? How was it able to transfer its heat to the popcorn inside?

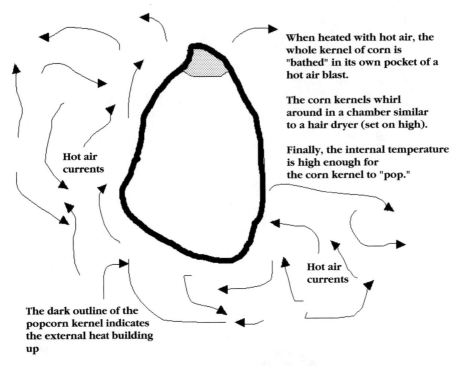

3-2 Popcorn convection

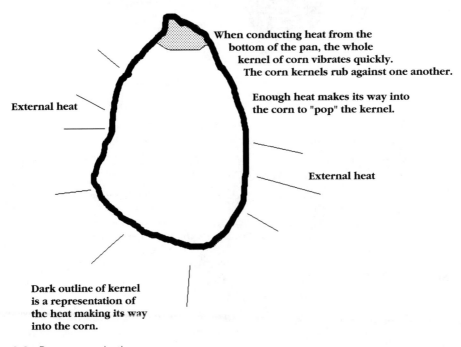

3-3 Popcorn conduction

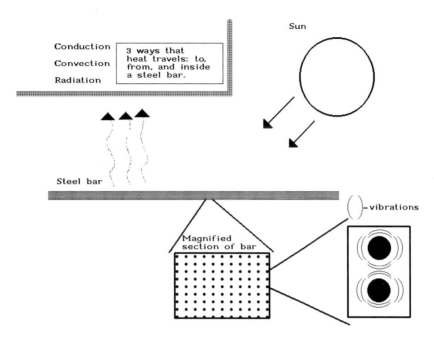

3-4 Conduction of a steel bar

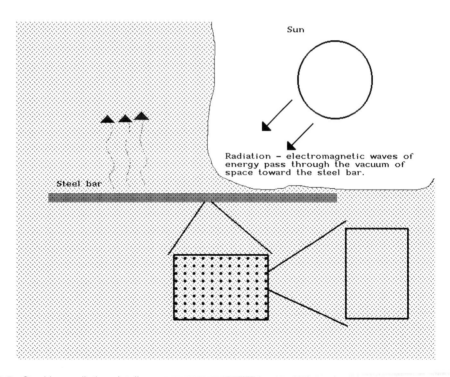

3-5 Steel bar radiation detail

Radiation, convection, & conduction 15

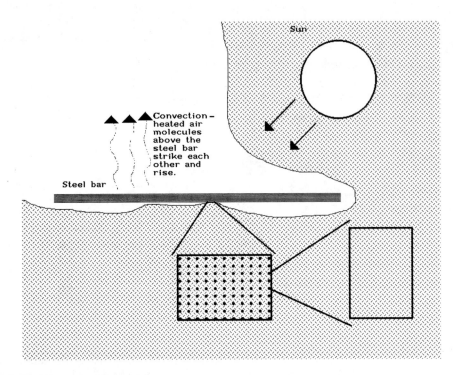

3-6 Steel bar convection detail

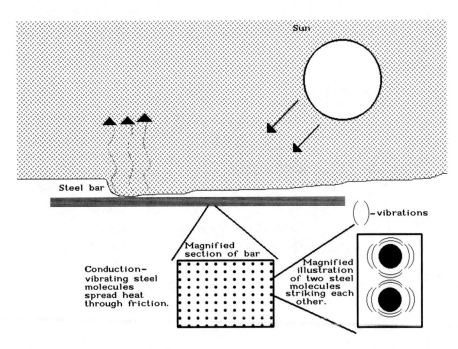

3-7 Steel bar conduction detail

QUESTIONS/CONCLUSIONS

1. How does the earth receive its heat from the sun?
2. How does a greenhouse (for plants) operate?
3. How are mirages formed?
4. What principle of warming does a blow-dryer use?
5. What principle of warming does a fireplace exhibit?
6. What might be the danger of a human body being exposed to microwave radiation as compared to being burned by a hot stove?

POST-LAB ACTIVITY—LAB 3

Popcorn has an extra-hard shell surrounding a larger amount of moisture than is found in regular corn. When this moisture is heated (by any of these three forms of energy) the steam produced is enough to "shatter" this shell and make a fluffy piece of popcorn.

Mass a sample of an appropriate number of grains of popcorn. Divide the mass by the number of corn kernels. This will give an average mass for one kernel.

Pop each sample of corn using the energy forms of convection, conduction, and radiation. After each sample has cooled off, mass the popped corn again. What mass difference (if any) is there? Which form of energy gives the biggest mass difference? What mass has the corn kernel lost/gained in being heated? Is the popping of popcorn a chemical or physical change? (It's most certainly a chemical change if the popcorn is burned! Don't reach that point during the heating of any popcorn sample.) Make copies of TABLE 3-1 for your own use.

Table 3-1
Convection, conduction, radiation

(circle type used)

A.	Number of kernels used	_____
B.	Mass of kernels before heating	_____ g
C.	Mass of one kernel (B / A)	_____ g
D.	Mass of popped corn after heating	_____ g
E.	Mass of one popped kernel (D / A)	_____ g
F.	Difference in before/after masses (E − C or C − E, positive result)	_____ g
G.	Was this a gain or a loss in mass? GAIN LOSS (circle one)	

Also see questions 1 and 9 in appendix C, "Environmentally Speaking."

4
Heat in the atmosphere

An object closer to a heat source will heat faster than an object farther away from the same heat source. This lab shows that the angle of radiation is also important. Objects at the north and south poles are heated more indirectly than objects at the equator. At the poles, there's more atmosphere for light rays to travel through than there is at the equator.

PURPOSE
The purpose of this lab is to observe the relationship between a heat source's temperature and distance from the heated object.

MATERIALS NEEDED
- Lamp with 100-watt bulb.
- Four thermometers.
- Meter stick.
- Tape.
- Watch or clock with second hand.
- Graph paper.
- Ruler.
- Pencil.

PROCEDURE
1. See FIG. 4-1.
2. Tape the thermometers onto the meterstick at 25, 50, 75, and 100 centimeters. Don't cover the thermometer bulbs.

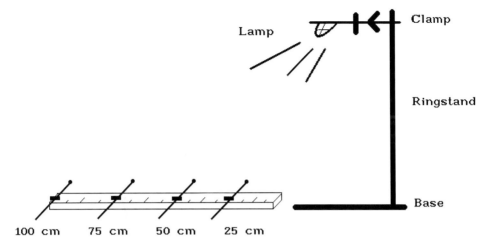

4-1 Lab setup

3. Record the beginning temperature for each thermometer on FIG. 4-2.
4. Turn on the lamp.
5. Record the heating temperatures from minute 1 to minute 9 on FIG. 4-3.
6. At minute 10, record the temperatures on FIG. 4-3. Turn off the lamp.
7. Record the cooling temperatures from minute 11 to minute 20 on FIG. 4-4.
8. See FIG. 4-5 for a sample graph of similarly recorded data.
9. Graph your recorded data on graph paper using an appropriate scale.

Beginning temperatures (deg. C)			
1	2	3	4
Thermometer #			

4-2 Beginning temperatures

Heat in the atmosphere

Time (minutes)	Heating Temperature (deg. C)			
	1	2	3	4
1				
2				
3				
4				
5				
6				
7				
8				
9				
10				
	Thermometer #			

4-3 Heating temperatures

OBSERVATIONS

1. Which thermometer received the strongest light?

QUESTIONS/CONCLUSIONS

1. In the Northern Hemisphere, the earth is closest to the sun in the winter, yet it's colder in winter than in summer. Why is this?
2. What feature of the earth is crucial to the change of seasons throughout the year?

20   The sun

4-4 Cooling temperatures

Time (minutes)	Cooling Temperature (deg. C)			
11				
12				
13				
14				
15				
16				
17				
18				
19				
20				
	1	2	3	4
	Thermometer #			

POST-LAB ACTIVITY—LAB 4

Perform the lab this time with the meterstick/thermometers in a glass aquarium. (If the aquarium is not of sufficient length, scale down the distances between the thermometers.) Seal off the aquarium with plastic wrap or a glass cover, or turn the aquarium upside down on a lab table, covering the thermometers.

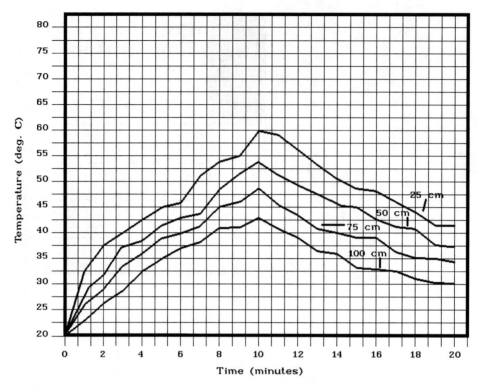

4-5 Example graph

Shine the light at the thermometers. (See FIG. 4-1 for arrangement.) Record the temperatures for the same time interval. Graph your results. How is this graph the same as the first one? How is it different? How is the glass of the aquarium affecting the rays of the heating light? How is this effect similar to a greenhouse? How is it different?

Also see questions 59 and 60 in appendix C, "Environmentally Speaking."

The Earth

Currently home to all living creatures, the earth is a sphere, somewhat flattened at the poles and bulging at the equator. This bulge is caused by the earth's spin. Our myriad of weather changes and patterns are due to the earth's rotation on its axis, revolution around the sun, and 23½-degree tilt.

One-half of the earth is always in darkness, the other half in light. It's these dark/light contrasts that set up uneven heating patterns on the surface of the planet. From the poles to the equator, the earth's surface plays host to intricate heating/cooling processes. Labs 5, 6, and 7 each explore different parts of this heating/cooling system.

5
Temperature & evaporation of water

When evaporation occurs, cooling takes place. Think of how cool you've felt getting out of a pool on a windy summer day. Water molecules were evaporating from your skin's surface, lowering its temperature as each water molecule "stripped" a portion of heat from your skin.

PURPOSE

The purpose of this lab is to demonstrate the evaporation rates of specific amounts of water that are different distances from a heat source.

MATERIALS NEEDED

- Lamp with 100-watt bulb.
- Five shallow bowls.
- 15-centimeter plastic ruler.
- Meterstick.
- Water.

PROCEDURE

Caution! Be sure electrical cords are coiled away from any sink or pans filled with water.

1. See FIG. 5-1.
2. Arrange the meterstick, lamp, and pans as shown in FIG. 5-1.
3. Fill the pans with exactly 2.0 centimeters of water.
4. Lay a thermometer over each pan.

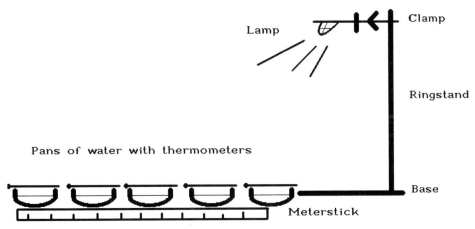

5-1 Lab setup

5. Turn on the lamp.
6. After 10 minutes, record the temperature of each thermometer on the chart shown in FIG. 5-2.
7. Leave the lamp on for 24 hours.
8. The next day, record the new height of the water to the nearest 0.1 centimeter on the chart shown in FIG. 5-2.

OBSERVATIONS

1. Would the evaporation amount have been the same if a different level of water had been used? Why or why not?
2. Would the rate of water evaporation have been different if the pan had been shaped differently?

Bowl #	Temperature (deg. C)	Initial height of water (cm)	New height of water (cm)	Amount of water evaporated (cm)
1		2.0		
2		2.0		
3		2.0		
4		2.0		
5		2.0		

5-2 Heat evaporation data table

QUESTIONS/CONCLUSIONS

1. Sometimes it rains heavily over the desert. Yet, the rain can evaporate before reaching the ground. Can you explain why?
2. Where on the earth would you expect the greatest rate of evaporation from sunlight? Why is this?
3. Define the term *albedo*.
4. How does the albedo of an object determine evaporation rate?

POST-LAB ACTIVITY—LAB 5

Perform the lab this time with different types of liquid in each container. Use salt water, tap water, distilled water, water/silt, and water/isopropyl alcohol. *(Wear goggles for safety.)* Be sure all levels are the same. See FIG. 5-1 for initial setup. Try different placements of the pans. (Instead of placing them in a line, center the light over the whole group.) Run this lab for 24 to 48 hours. Record your results. How are these results different from the original lab results? How does the location of each pan affect its own evaporation rate? Does the addition of impurities (salt, sugar, silt, and isopropyl alcohol) affect the evaporation? What additional tests could be done to further investigate this?

Also see questions 1 and 60 in appendix C, "Environmentally Speaking."

6
Heating by convection

This experiment features a model of wind traveling from a warmer area to a cooler area. (See FIG. 6-1 for an illustration of the cycle caused by heating and cooling.) There are many examples of these kinds of cycles throughout a given region. Can you name some of these cycles?

PURPOSE
The purpose of this lab is to observe a convection cycle of heated/cooled air.

MATERIALS NEEDED
- Lamp with 60-watt or 75-watt bulb.
- Large aquarium.
- Nine thermometers.
- Watch or clock with second hand.
- Masking tape.
- One 500-milliliter beaker filled with ice cubes.
- Plastic wrap or glass cover.

PROCEDURE
Caution! Be sure electrical cords are coiled away from any sink or pans filled with water.

1. See FIG. 6-2.
2. Tape the nine thermometers as shown in FIG. 6-2. Be sure the readings are visible from the outside of the aquarium.

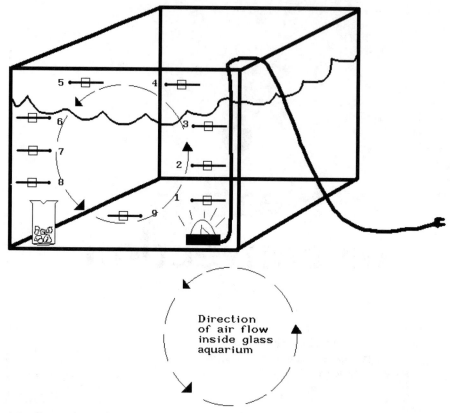

6-1 Convection cycle

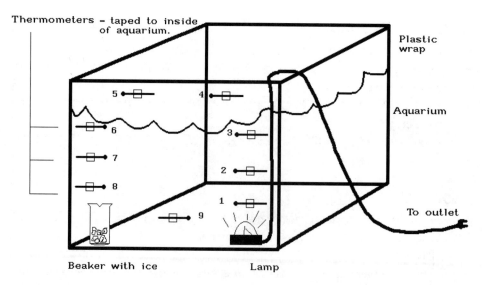

6-2 Lab setup

3. Wait two minutes. Record beginning temperatures on FIG. 6-3.
4. Place the lamp in one end of the aquarium.
5. Place the beaker with ice cubes at the opposite end.
6. Cover the aquarium with plastic wrap or a glass cover.
7. Turn on the lamp to begin heating.
8. After five minutes, record the temperatures on FIG. 6-3.
9. After 10 more minutes, turn the lamp off. On FIG. 6-3, record the temperature of each thermometer.
10. See FIG. 6-4 for a sample graph of similarly recorded data.
11. Graph the recorded results using graph paper and the appropriate scale.

OBSERVATIONS

1. How quickly did the temperatures change?
2. Which temperatures were the coolest, and which were the warmest?
3. Did the temperatures plateau (level off?)

Thermometer #	TEMPERATURE (deg. C)		
	Beginning readings	Readings after 5 min.	Readings after 10 min.
1			
2			
3			
4			
5			
6			
7			
8			
9			

6-3 Data table

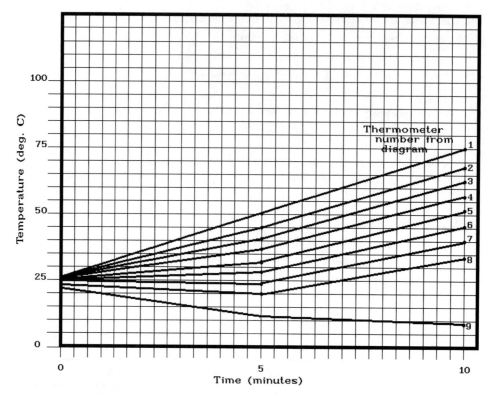

6-4 Example graph of data

QUESTIONS/CONCLUSIONS

1. Define the term *convection cycle*.
2. How is a convection cycle responsible for a change in weather?
3. Define the term *microweather*. Why is microweather so difficult to predict?

POST-LAB ACTIVITY—LAB 6

Perform the lab again, but at minute number 10, switch the locations of the lamp and the beaker of ice. (Put the lamp where the ice was, and put the ice where the lamp was. See FIG. 6-2 again.) After five minutes, record the temperatures. Record the temperatures again at 10 minutes.

Graph this data. Describe the trends of each of the thermometer readings. How have any of them changed? What caused the reversal in readings? Describe the direction of the "wind flow" inside the aquarium. How is it different from the results from the initial lab activity?

Also see questions 1, 59, and 60 in appendix C, "Environmentally Speaking."

7
Determining dew point

Figure 7-1 shows the "microweather" of saturated air surrounding a beaker. This is similar to a valley with trapped, moist air surrounded by hills. An early morning fog could eventually cover the region when this moist air cools.

PURPOSE
The purpose of this lab is to calculate the dew point of air inside a container filled with ice. (*Dew point* is the temperature at which water vapor in the air begins to condense.)

MATERIALS NEEDED
- One thermometer.
- One 500-milliliter beaker.
- Ice.

PROCEDURE
1. See FIG. 7-2.
2. On FIG. 7-3, record the temperature of the air just inside the opening of the empty 500-milliliter beaker.
3. Hold the thermometer just inside the opening of the beaker for the next two steps.
4. Fill the beaker with several ice cubes. Don't let the thermometer bulb touch the ice cubes. See FIG. 7-4.
5. When drops of moisture appear on the outside of the container, record the temperature on FIG. 7-3. This is the dew point.
6. See FIGS. 7-5 and 7-6 to calculate the relative humidity of the air above the ice cubes.

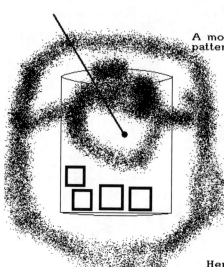

A model of the beaker's own microweather pattern with ice cubes.

Here, the spray pattern is a representation of the effects of the cold from the ice cubes.

Depending on the amount of water vapor surrounding the beaker, condensation will form on the exterior of the glass surface.

7-1 Microweather model

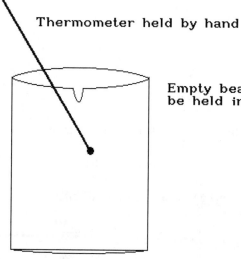

Thermometer held by hand

Empty beaker – the thermometer must be held inside the dry beaker.

7-2 Lab setup

Empty beaker temperature (deg. C) A	Temperature when condensation forms (deg. C) B	Difference in both readings (A-B)	Relative humidty (%)

7-3 Data table

Beaker with ice cubes — the thermometer must be held just inside the beaker, but above the ice. When condensation forms on the beaker's exterior, record that temperature.

ice cubes

7-4 Beaker with cubes

OBSERVATIONS

1. Why did you have to wait for condensation to form?
2. When condensation formed, was the air around the beaker saturated?

QUESTIONS/CONCLUSIONS

1. Define the term *lifting condensation level.*
2. How does the lifting condensation level relate to cloud formation?
3. Define the terms *ground fog* and *radiation fog.*
4. What's the relative humidity of the air in a ground or radiation fog?

Determining dew point 33

Empty beaker temperature (deg. C)	Temperature difference (deg. C)									
	1	2	3	4	5	6	7	8	9	10
6	86	73	60	48	35	24	11			
8	87	75	63	51	40	29	19	8		
10	88	77	66	55	44	34	24	15	6	
12	89	78	68	58	48	39	29	21	12	
14	90	79	70	60	51	42	34	26	18	10
16	90	81	71	63	54	46	38	30	23	15

RELATIVE HUMIDITY AROUND BEAKER

7-5 Humidity table I

POST-LAB ACTIVITY—LAB 7

Determine the dew point of the air inside your bathroom at home. Record the temperature inside your bathroom with the door closed at the highest point you can reach on the mirror. Now, run the shower at a fairly warm (not hot) temperature. When water condenses on the mirror (becomes fogged), record the temperature. Be sure the thermometer is at the same level as the initial reading. Using the data table used in this experiment, determine the dew point. Does size of the bathroom affect the dew point of the room?

Also, try measuring the dew point under the level where the condensation forms on the mirror. Take thermometer readings near the floor before and after the condensation forms. (You might have to do this on separate days to allow the bathroom to "dry out.") How different is this to the dew point above the condensation line?

Also see question 60 in appendix C, "Environmentally Speaking."

Empty beaker temperature (deg. C)	Temperature difference (deg. C)									
	1	2	3	4	5	6	7	8	9	10
18	91	82	73	65	57	49	41	34	27	20
20	91	83	74	66	59	51	44	38	31	24
22	92	83	76	68	61	54	47	41	34	28
24	92	84	77	69	62	56	49	44	37	31
26	92	85	78	71	64	58	51	47	40	34
28	93	85	78	72	65	59	53	48	42	37

RELATIVE HUMIDITY AROUND BEAKER

7-6 Humidity table 2

PART 3

The Atmosphere

The third most important weather category is the atmosphere itself. It's composed of many gases as well as water, smoke, dust, and salt particles. The atmosphere has a total weight of nearly 6,000 trillion tons. Likened to the skin of an apple in relative thickness, our atmosphere is held to the earth by gravity. Nearly all weather phenomena occur at the bottom eight miles of the atmosphere.

It's water vapor that plays the most important role in the atmosphere. Water can take three different forms: vapor, liquid, and ice. If all the water in the earth's atmosphere were to condense at once, it would cover the surface of the planet to a depth of nearly one inch. The next 11 labs are a selection of "make-and-take" projects, demonstrations, and group work that attempt to explain some of the intricate behavior of our atmosphere.

8
The barometer

In this lab, you'll make a simple barometer that can detect when a low-pressure or high-pressure front has passed through the region.

PURPOSE
The purpose of this lab is to make a barometer to measure fluctuations in air pressure.

MATERIALS NEEDED
- Large glass jar.
- Large, uninflated balloon.
- Rubber band.
- Masking tape.
- Toothpick.
- 3-inch-by-5-inch card.
- White glue.
- Pencil.

PROCEDURE
1. See FIG. 8-1.
2. Be sure the jar is clean and dry.
3. Cut off the neck of the balloon. Discard the neck.
4. Evenly stretch the remainder of the balloon over the jar.
5. Secure with a rubber band.

38 The atmosphere

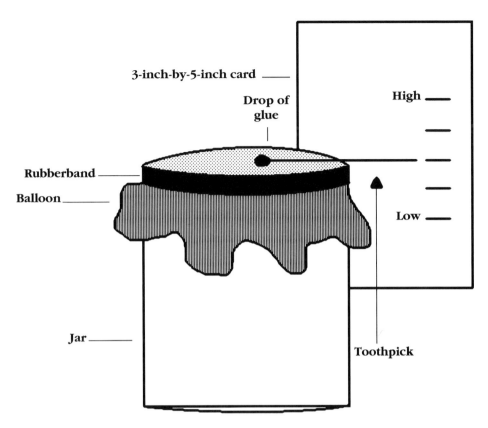

8-1 Construction model

6. Glue the toothpick onto the balloon.
7. With tape, mount the card on the side of the jar.
8. Above the toothpick, write the word "High."
9. Below the toothpick, write the word "Low."
10. Place the barometer indoors away from moisture, heat, or windows. Be sure to keep the apparatus at a constant temperature.
11. Observe the barometer daily for any changes. See FIGS. 8-2 and 8-3 for examples.

OBSERVATIONS

1. When your barometer pointed to Low, what was the weather like within one to two days?

QUESTIONS/CONCLUSIONS

1. What are the differences between an *aneroid* and a *mercurial* barometer?
2. Which type of the preceding barometers is more accurate? Why?

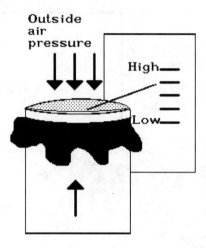

High pressure

High pressure exists when the outside air pressure is greater than the pressure inside the glass jar.

8-2 Barometer high reading

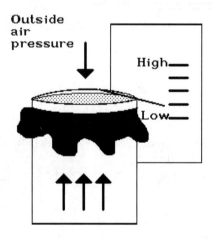

Low pressure

Low pressure exists when the outside air pressure is less than the pressure inside the glass jar.

8-3 Barometer low reading

POST-LAB ACTIVITY—LAB 8

How will heat and cold affect the reading of this barometer? Place it in a refrigerator. Wait approximately 30 minutes. Remove the barometer and place it in a pan of hot water (approximately 35 degrees Celsius, 95 degrees Fahrenheit).

How quickly did the pressure change? What's responsible for the fluctuations? If the interior of the glass chamber were a partial vacuum, how would this affect the direction of the indicator? Would the readings be more accurate? Why did the experiment call for a glass container and not a metal can?

9
Calculating the size of a raindrop

Raindrops reach a maximum size before breaking into smaller drops. Also, contrary to popular belief, falling rain is not teardrop shaped. See FIG. 9-1 for examples. In this lab, you'll calculate raindrop size.

```
Shapes of falling raindrops:    (Not actual size in these drawings.)

At diameters of .08 inches or less, a falling drop looks like this:
                    ○  (larger sphere)

Larger than this, but less than a 1/4-inch in diameter, the drop bulges:
   The air pressure
   flattens the bottom.   ⌒

At diameters greater than 1/4 inch, the droplet breaks up:
                    ○ ○  (two or more smaller spheres)
```

9-1 Shapes of falling drops

PURPOSE

The purpose of this lab is to "capture" several raindrops and calculate their volumes.

MATERIALS NEEDED

- Old nylon hose.
- Rubber band.
- Wide-mouthed glass jar.
- Flour.
- Centimeter ruler.
- Pencil.
- Calculator.
- Rainy day.

PROCEDURE

1. See FIG. 9-2.
2. Cut a 15-centimeter-by-15-centimeter square into the nylon hose.
3. Stretch the nylon tightly over the jar's mouth.
4. With a rubber band, securely attach the nylon to the glass jar.
5. Dust the hose evenly and lightly with a layer of flour.

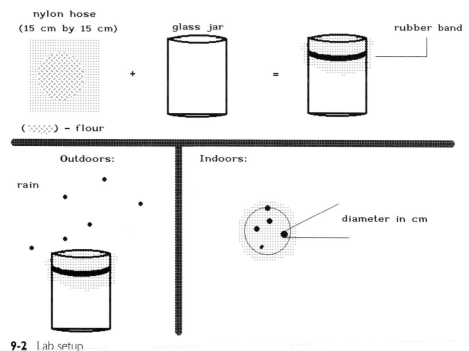

9-2 Lab setup

6. Go outdoors (in the rain).
7. Quickly expose the nylon to the rain to capture several drops.
8. Go indoors.
9. Measure the diameter of three of the drops. Record these diameters on FIG. 9-3.
10. See FIG. 9-4 to calculate the volume of a drop.

Drop number	Diameter (cm)	Radius (cm)	Volume (cc = ml)
1			
2			
3			

9-3 Raindrop size data table

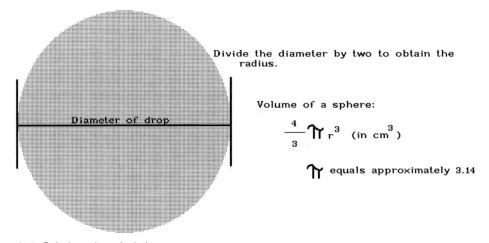

Divide the diameter by two to obtain the radius.

Volume of a sphere:

$$\frac{4}{3} \pi r^3 \quad (\text{in cm}^3)$$

π equals approximately 3.14

9-4 Raindrop size calculations

OBSERVATIONS

1. What were the shapes of the impressions left by the raindrops after they struck the nylon hose? Why were they shaped like this?

QUESTIONS/CONCLUSIONS

1. What are the approximate diameters of condensation nuclei, fog, and drizzle?
2. How do raindrops become so large as they fall?
3. Describe the two theories that account for rain formation: collision/coalescence, and ice crystal.
4. What do all raindrops need in order to become falling drops?
5. How does smoke assist in the formation of rain?

POST-LAB ACTIVITY—LAB 9

Obtain mineral oil, cooking oil, or glycerin. Pour 200 milliliters in a plastic see-through container. Now, "trap" several raindrops in this container. Describe the shapes of the drops as they strike the surface. Watch as the drops travel to the bottom of the container. What form does the droplet appear to take now?

Relate the raindrops falling into the cup with the rain striking a dry highway at the beginning of a rainy season. More accidents occur just after the beginning of a light rain. Why is this? What possible contaminants are covering the surface of the roadway? What have various transportation agencies done to the road's surface to prevent it from being so slippery?

☆ ☆ **FACTOIDS** ☆ ☆

The following facts were paraphrased from the 1992 *WORLD ALMANAC AND BOOK OF FACTS*:

- An inch of rain? Spread over one acre, this inch of water would weigh more than 200 thousand pounds. That's nearly 30 thousand gallons of water.
- According to records kept since 1850 by the British Meteorological Office, 1990 was the warmest year for the planet. The average global temperature for 1990 was nearly 60 degrees Fahrenheit.
- The amount of oil spilled into the Northern Arabian Gulf (during the Persian Gulf War) has been estimated to be as much as 130 million gallons.

☆ ☆ ☆ ☆ ☆ ☆ ☆ ☆ ☆ ☆ ☆ ☆ ☆ ☆ ☆ ☆ ☆ ☆ ☆

10
The relative humidity indicator

The indicator you make in this lab will be able to detect when an air mass surrounding you is moist or dry. Figures 10-1 through 10-4 show how liquid water evaporates and then condenses in a heated/cooled chamber. The air in this hypothetical, sealed aquarium is first unsaturated, then becomes quite saturated out in the sunlight.

PURPOSE
The purpose of this lab is to observe changes in the moisture of the outside air.

MATERIALS NEEDED
- Cobalt chloride ($CoCl_2$) crystals.
- Salt.
- Distilled water.
- Untreated white cotton cloth.
- Black poster board.
- Glue stick or white glue.

PROCEDURE
Caution: Wear plastic gloves and safety goggles at all times when handling the cobalt chloride solution. Practice strict hygiene when using this substance. It's poisonous!

1. Dissolve as much cobalt chloride as possible in 100 milliliters of distilled water.
2. Add 20 grams of salt to the cobalt chloride solution. Stir to dissolve.
3. Soak several 4-centimeter-by-4-centimeter cotton squares in the solution.

46   The atmosphere

4. Take the squares out and let them dry in the sun or under a heat lamp.
5. With the glue stick or very little white glue, mount one dry square on a black 10-centimeter-by-10-centimeter cardboard square.
6. After handling the cobalt chloride solution, wash your hands thoroughly.
7. Dispose of the remaining solution according to your instructor or according to your lab safety procedures. (The cobalt chloride can be washed down the sink with plenty of cold water—or store the chemical in an appropriately labeled container for future use.)
8. Place the relative humidity indicator indoors away from moisture, heat, or windows.
9. Observe your indicator daily for any changes. See FIG. 10-5. Note: It's possible for your indicator to show pink/red and have no precipitation follow.

Glass chamber

♥ - enlarged view of a water molecule

The percentage of relative humidity is the mass of water vapor in the air divided by the highest amount of water vapor that the air can hold at a given temperature, multiplied by 100.

The percentage can range from zero percent (0%)— very dry—to one-hundred percent (100%)—rainy or very humid.

If moisture enters or leaves the air, or if the temperature varies, the relative humidity also varies.

This chamber has perfectly dry air. A container of water placed inside of it has the same temperature as the air.

Immediately, water molecules leave the liquid phase and bounce around inside the chamber as water vapor.

10-1 Humidity series I

OBSERVATIONS

1. What color was your indicator just before a storm?
2. After a rain or snow, how long did it take the humidity indicator to change colors?

QUESTIONS/CONCLUSIONS

1. How does a hygrometer (humidity indicator) work?
2. Research the operation of a sling psychrometer. How does it work?
3. After physically exerting yourself, why are you more uncomfortable in a classroom at room temperature on a rainy day than you would be in a classroom at the same temperature on a hot, dry, summer day?

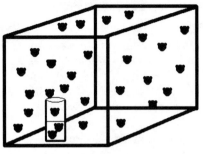

Here, water molecules are going back and forth from the liquid phase to the gaseous phase. When the number of molecules leaving the water equals the number of molecules re-entering it, the air in the chamber is "saturated."

♥ - enlarged view of a water molecule.

10-2 Humidity series 2

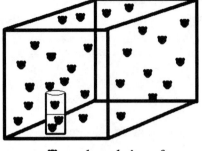

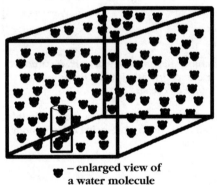

The chamber is placed in the sun. As the temperatures of both the air and the water rise, more water molecules leave the liquid phase and become water vapor.

The air in the chamber is now quite saturated and will remain so if the sun continues to shine on the chamber.

♥ – enlarged view of a water molecule

10-3 Humidity series 3

POST-LAB ACTIVITY—LAB 10

Wearing gloves and safety goggles, tape a piece of cobalt chloride-soaked cloth to the interior bottom of a large, wide-mouthed jar. (Wash your hands after handling this cloth.) Fill a small (50-milliliter) beaker with 20 milliliters of water. Place the beaker on the interior of the lid. With the jar upside down, carefully twist the jar onto the lid. Hold the lid while turning the jar. Place the apparatus in sunlight or under a heating lamp. How quickly did the indicator show moisture changes?

Take the beaker of water out and stand the jar upright in the sun or under a heating lamp. Did the indicator change colors again? Why is this indicator not accurate enough to detect any small changes in humidity?

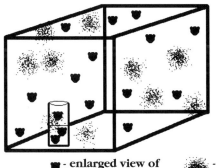

The chamber has now been brought indoors.

Assume the temperature indoors is quite cool. Almost immediately, the high-speed water vapor molecules slow down. As they slow down, they strike other slow-moving molecules.

In just several minutes, condensation will form throughout the chamber on the interior walls.

Some water vapor molecules will stay as vapor in the saturated air.

♥ - enlarged view of a water molecule   ✹ - condensation

10-4 Humidity series 4

BLUE	BLUE/PINK	PINK
fair	change	stormy

Note: color changes can take a period of several days to complete.

A fog, early morning dew, or even a wind blowing the spray from a garden sprinkler can cause a change from blue to pink.

10-5 Humidity indicator table

11 The wind chill factor

When water evaporates, cooling takes place and temperature decreases. Quite the opposite is true when condensation occurs. Condensation is a warming process, and heat is given off when gaseous water returns to the liquid phase. In this lab, you'll see a model of what happens to the temperature when a wind blows across a body of water.

PURPOSE
The purpose of this lab is to observe a model that demonstrates evaporation and wind chill.

MATERIALS NEEDED
- Shallow pan.
- Water.
- Thermometer.
- Fan.
- Watch or clock with second hand.
- Graph paper.
- Pencil.

PROCEDURE
1. See FIG. 11-1.
2. Place 3 centimeters of water in the pan.
3. Place the thermometer in the pan.

4. Make sure the bulb of the thermometer is partially submerged.
5. Record the beginning temperature on FIG. 11-2.
6. Start the fan.
7. On FIG. 11-2, record the temperature every minute until minute 10.
8. See FIG. 11-3 for a sample graph of similarly recorded data.
9. Graph recorded data on graph paper using the appropriate scales.
10. See FIGS. 11-4 through 11-10 for wind chill charts to use for outside temperatures and wind speeds for both Fahrenheit and Celsius. You can obtain the wind speed from a radio broadcast or by an anemometer (if available).

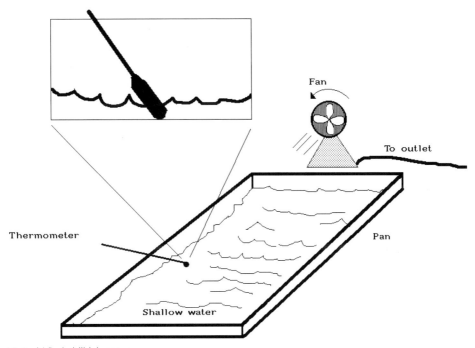

11-1 Wind chill lab setup

Minute	0	1	2	3	4	5	6	7	8	9	10
Temperature (deg. C)											

11-2 Wind chill data table

The wind chill factor 51

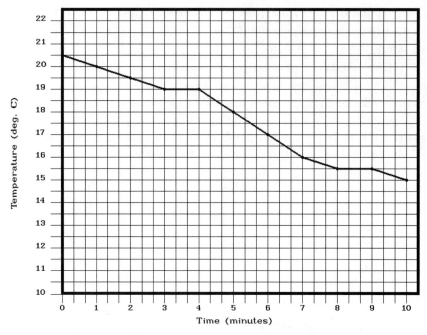

11-3 Wind chill example graph

Actual temperature (deg. F)	Wind speed (miles per hour)							
	5	10	15	20	25	30	35	40
35	33	21	16	12	7	5	3	1
34	32	20	15	10	6	4	2	0
33	31	19	14	8	4	2	0	-1
32	29	18	13	7	3	1	-1	-2
31	28	17	12	5	1	-1	-3	-3
30	27	16	11	3	0	-2	-4	-4
29	26	15	9	2	-1	-4	-6	-6
28	25	13	7	0	-3	-6	-8	-8
27	23	12	5	-1	-4	-7	-9	-10
26	22	10	3	-3	-6	-9	-11	-13
25	21	9	1	-4	-7	-11	-13	-15
24	20	8	0	-5	-9	-12	-14	-16
23	19	6	-2	-6	-10	-14	-16	-18

Equivalent temperature

11-4 Wind chill Fahrenheit chart 1

Actual temperature (deg. F)	Wind speed (miles per hour)							
	5	10	15	20	25	30	35	40
22	18	5	-3	-7	-12	-15	-17	-19
21	17	3	-5	-8	-13	-17	-19	-21
20	16	2	-6	-9	-15	-18	-20	-22
19	15	1	-7	-11	-16	-20	-21	-23
18	14	0	-8	-12	-18	-21	-23	-25
17	14	0	-9	-14	-19	-23	-24	-26
16	13	-1	-10	-15	-21	-24	-26	-28
15	12	-2	-11	-17	-22	-26	-27	-29
14	11	-3	-12	-18	-23	-27	-29	-30
13	10	-5	-14	-20	-25	-29	-30	-32
12	9	-6	-15	-21	-26	-30	-32	-33
11	8	-8	-17	-23	-28	-32	-33	-35
10	7	-9	-18	-24	-29	-33	-35	-36

Equivalent temperature

11-5 Wind chill Fahrenheit chart 2

Actual temperature (deg. F)	Wind speed (miles per hour)							
	5	10	15	20	25	30	35	40
9	6	-10	-19	-26	-31	-35	-37	-38
8	5	-11	-21	-27	-32	-36	-38	-40
7	3	-13	-22	-29	-34	-38	-40	-41
6	2	-14	-24	-30	-35	-39	-41	-43
5	1	-15	-25	-32	-37	-41	-43	-45
4	0	-16	-27	-34	-39	-43	-45	-48
3	-2	-18	-28	-35	-40	-44	-47	-49
2	-3	-19	-30	-37	-42	-46	-48	-50
1	-5	-21	-31	-38	-43	-47	-50	-52
0	-6	-22	-33	-40	-45	-49	-52	-54
-1	-7	-23	-34	-41	-46	-50	-54	-56
-2	-8	-24	-36	-42	-48	-52	-55	-57
-3	-9	-25	-37	-44	-49	-53	-57	-59
-4	-10	-26	-39	-45	-51	-55	-58	-60

Equivalent temperature

11-6 Wind chill Fahrenheit chart 3

Actual temperature (deg. F)	Wind speed (miles per hour)							
	5	10	15	20	25	30	35	40
-5	-11	-27	-40	-46	-52	-56	-60	-62
-6	-12	-28	-41	-47	-53	-57	-61	-63
-7	-13	-29	-42	-48	-54	-59	-63	-65
-8	-13	-29	-43	-50	-56	-60	-64	-66
-9	-14	-30	-44	-51	-57	-62	-66	-68
-10	-15	-31	-45	-52	-58	-63	-67	-69

Equivalent temperature

11-7 Wind chill Fahrenheit chart 4

Actual temperature (deg. C)	Wind speed (kilometers per hour)							
	Calm	10	20	30	40	50	60	70
10	10	8	3	1	-1	-2	-3	-4
5	5	2	-3	-6	-8	-10	-11	-12
0	0	-3	-10	-13	-16	-18	-19	-20
-5	-5	-9	-16	-20	-23	-25	-27	-28
-10	-10	-14	-23	-27	-31	-33	-35	-35

Equivalent temperature

11-8 Wind chill Celsius chart 1

Actual temperature (deg. C)	Wind speed (kilometers per hour)							
	Calm	10	20	30	40	50	60	70
-15	-15	-20	-29	-34	-38	-41	-42	-43
-20	-20	-25	-35	-42	-46	-48	-50	-51
-25	-25	-31	-42	-49	-53	-56	-58	-59
-30	-30	-37	-48	-56	-60	-64	-66	-67

Equivalent temperature

11-9 Wind chill Celsius chart 2

The atmosphere

Actual temperature (deg. C)	Wind speed (kilometers per hour)							
	Calm	10	20	30	40	50	60	70
−35	−35	−42	−55	−63	−68	−71	−74	−75
−40	−40	−48	−61	−70	−75	−79	−82	−83
−45	−45	−53	−68	−77	−83	−87	−90	−91
−50	−50	−59	−74	−84	−90	−94	−97	−99

Equivalent temperature

11-10 Wind chill Celsius chart 3

OBSERVATIONS

1. Did the temperature appear to drop steadily?
2. Was there any plateau (leveling off) in the temperature as it dropped?
3. At what minute did the temperature drop and remain at until the end of 10 minutes?

QUESTIONS/CONCLUSIONS

1. If hot water (instead of tap water) had been used, what would the results have been?
2. As you climb out of a pool on a warm day, you feel colder. Explain why this occurs in terms of evaporation.

POST-LAB ACTIVITY—LAB 11

Perform this lab again, but use isopropyl alcohol, or a water/isopropyl alcohol mixture in the pan. *(Be sure to wear safety goggles. After use, the alcohol/water mixture can be washed down the sink.)* Also, try salt or distilled water. Is there any difference in the temperature recorded?

Graph the results. How is this graph similar or different from the one you originally made? How does the addition of impurities affect the evaporation of the liquid?

The wind chill factor 55

12
The weight of the atmosphere

We live at the bottom of a "sea" of air. As we move about in the atmosphere, it exerts a crushing 10 to 20 tons of pressure on each one of us. We don't feel it because our bodies push back with the same force.

PURPOSE
The purpose of this lab is to demonstrate the weight of the atmosphere on a compressible object.

MATERIALS NEEDED
- 2-liter plastic soft-drink bottle.
- Screw-on cap.
- Very hot water.

PROCEDURE

1. See FIG. 12-1.
2. Fill the container completely with very hot (not boiling) water.
3. Pour out the water into the sink. (The fastest way to drain this bottle is by swirling the contents with the container's mouth pointed into the sink.)
4. Immediately twist the cap tightly onto bottle.
5. Put the bottle on the desk.
6. Watch the action!

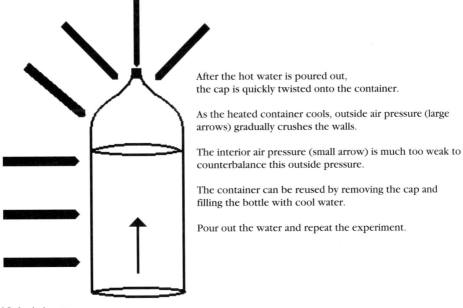

After the hot water is poured out, the cap is quickly twisted onto the container.

As the heated container cools, outside air pressure (large arrows) gradually crushes the walls.

The interior air pressure (small arrow) is much too weak to counterbalance this outside pressure.

The container can be reused by removing the cap and filling the bottle with cool water.

Pour out the water and repeat the experiment.

12-1 Lab setup

OBSERVATIONS

1. What happened to the container after you poured out the hot water and twisted the cap on?
2. How could you "re-inflate" the container?
3. What would you expect the air pressure inside the container to be after contraction?

POST-LAB ACTIVITY—LAB 12

Try performing this experiment under water. When the hot water is poured from the bottle, twist the cap on quickly. Submerge the bottle in three different water baths: ice water, tap water, and warm water. Is there any difference in how quickly the bottle becomes deformed in any particular water bath?

13
A convection cycle

Winds occur because of the uneven heating and cooling of regions. This lab will make a wind's path visible.

PURPOSE
The purpose of this lab is to observe a moving wind with smoke in heated air.

MATERIALS NEEDED
- Candle.
- Smoke chamber. (You can make a smoke chamber by covering a glass/plastic aquarium and having plastic PVC or glass tubes at opposite ends for the smoke to travel through.)
- Newspaper.
- Matches.

PROCEDURE
Follow all lab safety procedures when working with and extinguishing a flame. Also use caution with PVC pipe around an open file.

1. See FIG. 13-1.
2. Light the candle, place it in the chamber, and slide the window shut.
3. Light a rolled newspaper. Hold it over the chimney.
4. Watch what occurs.
5. When finished, dunk the smoldering newspaper in a pan of water.

OBSERVATIONS
1. In what direction does the smoke travel in the chamber?
2. What is the purpose of the lit candle? (Hint: see FIGS. 13-2 and 13-3.)

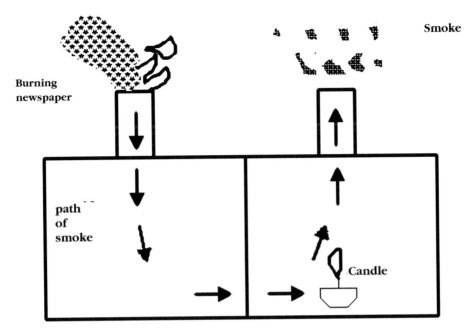

13-1 Lab setup

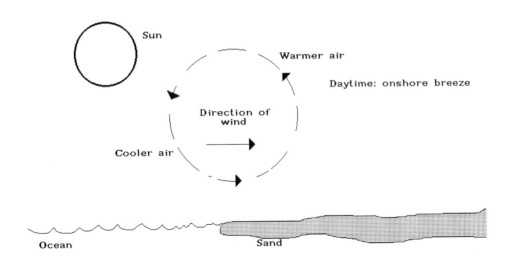

Land heats up more quickly than water; heat rises from the sand, allowing cooler air from over the ocean to move toward the shore.

13-2 Onshore breeze

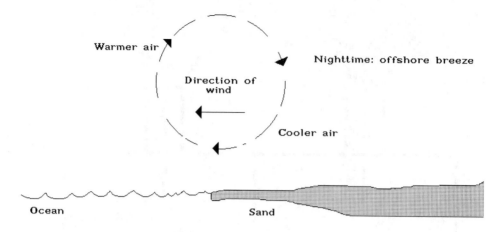

Land cools more quickly than water; cool air from the sand replaces the warmer air rising from the water. A breeze moves from the shore to the sea.

13-3 Offshore breeze

QUESTIONS/CONCLUSIONS

1. Define the term *Hadley Cell*. How is your convection cycle similar to a Hadley Cell?
2. How are winds formed?
3. How are high-pressure and low-pressure areas formed? Where are the high-pressures and low-pressures in this smoke chamber?
4. In what direction do winds flow from high or low pressure areas?
5. If you're at the beach on a sunny day, in what direction would the wind lift a kite at noon as opposed to early evening?

POST-LAB ACTIVITY—LAB 13

Perform this experiment again, but use a small beaker of crushed ice or water placed anywhere in the smoke chamber. Is there any difference in the direction of the smoke?

Try narrowing either "chimney." Use duct tape to constrict the holes. What effect does this have on the speed and path of the smoke entering and exiting the chamber?

Instead of using a candle inside the chamber, try a heating lamp outside the chamber. (*Caution! The lamp will be hot.*) Place it close enough to the exit chimney. What effect does an external heat source have on the direction of the smoke entering and exiting the chamber?

Also see questions 1 and 59 in appendix C, "Environmentally Speaking."

14
The cold front

A *cold front* is a shock wave of cool, dense air that slams into warm air, wedging it skyward. The warm air expands, cools, and condenses into a cloud formation called a *squall line*. The weather accompanying this squall line can be quite violent.

PURPOSE

The purpose of this lab is to observe a "collision" of cold and warm air, and the results of such a collision.

MATERIALS NEEDED

- Aquarium.
- Four sand bags.
- One 500-milliliter beaker.
- Hot water.
- Plastic wrap or glass covering.

PROCEDURE

1. See FIG. 14-1.
2. Be sure the sand bags have been in the refrigerator for at least three hours.
3. Place the sand bags in one corner of the aquarium.
4. Fill a 500-milliliter beaker with 400 milliliters of hot water.
5. Place the beaker in the corner of the aquarium, opposite the sand bags.
6. Cover the aquarium with plastic wrap or a glass cover.
7. Watch what occurs. See FIGS. 14-2 through 14-4.

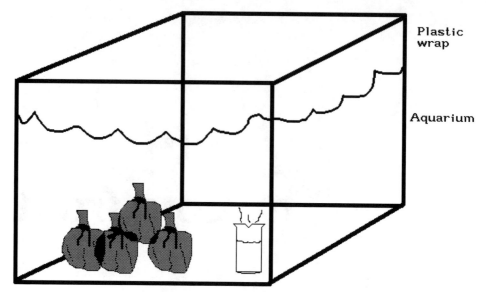

14-1 Lab setup

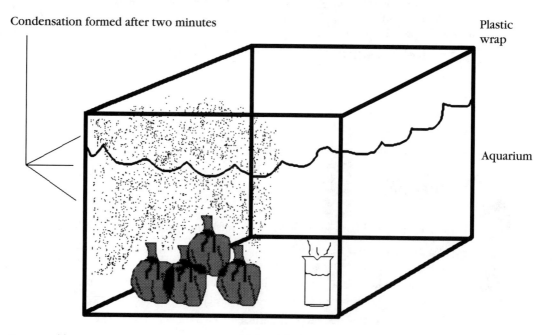

14-2 Example results

62 The atmosphere

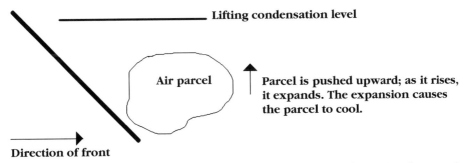

14-3 Cold front movement

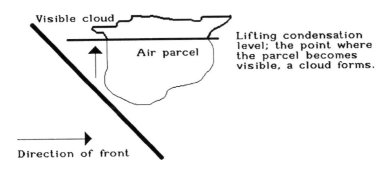

14-4 Cold front movement/cloud formation

OBSERVATIONS

1. Where was condensation forming in the aquarium?
2. Where did the evaporation occur in the aquarium?

QUESTIONS/CONCLUSIONS

1. How does a cold front form and move?
2. What happens when a cold front strikes warm air? What clouds form?

POST-LAB ACTIVITY—LAB 14

Obtain synoptic weather maps that show a cold front approaching your area. Watch TV or listen to the radio for information that indicates the approach of such a front. Just prior to its arrival in your region, carefully monitor the pressure, temperature, humidity, wind direction, and cloud type. Make copies of Labs 22, 23, and 24 for worksheets.

As the front passes through your area, record its "vital signs," the above-mentioned weather data. How does this compare to the trends of these signs before a cold front's arrival? How do they change as the front passes through your area?

15
The occluded front

When colder air "chases" and overtakes warmer air, the cold air lifts the warm air off the ground and sandwiches it between two cooler regions. When this low (composed of the lifted warm air) weakens, the two cooler regions mix with the low and form clouds.

PURPOSE

The purpose of this lab is to observe a "standoff" between sandwiches of cold and warm air.

MATERIALS NEEDED

- Aquarium.
- Four sand bags.
- One 500-milliliter beaker.
- Very hot water.
- Plastic wrap or glass covering.

PROCEDURE

1. See FIG. 15-1.
2. Be sure the sand bags have been in the refrigerator for at least three hours.
3. Place two sand bags in one corner of the aquarium.
4. Place the other two sand bags in a corner opposite the first two bags.
5. Fill a 500-milliliter beaker with 400 milliliters of hot water.
6. Place the beaker in the middle of the aquarium.

7. Cover the aquarium with plastic wrap.
8. Watch what occurs. See FIG. 15-2.

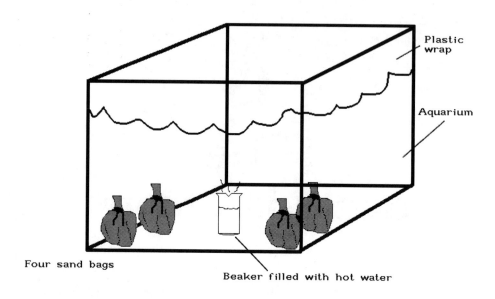

15-1 Lab setup

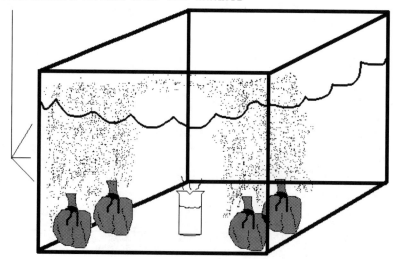

15-2 Example results

66 The atmosphere

OBSERVATIONS

1. Where did condensation form? Did it move in any direction in the aquarium?

QUESTIONS/CONCLUSIONS

1. How does an occluded front form?
2. What might be the weather at an occluded front? What clouds form?

POST-LAB ACTIVITY—LAB 15

Refer to the post-lab activity of Lab 14, and make the same observations of an occluded front as it approaches your region. Describe the trends of all data recorded.

16 The temperature inversion

Familiar to nearly all of us, smog is the result of a temperature inversion. Think of any large metropolitan city that has a smog problem, such as the San Fernando Valley in Los Angeles, or Manhattan, New York. These inversions are the result of moist, maritime air moving in from the sea. Pollutants then become trapped by the warm air and make it quite challenging to breathe.

PURPOSE

The purpose of this lab is to observe how warm air traps smoke particles over a layer of cool air.

MATERIALS NEEDED

- Aquarium.
- Lamp with 75-watt bulb.
- Six sand bags.
- Newspaper.
- One 500-milliliter beaker.
- Matches.
- Ringstand.
- Clamp.
- Plastic wrap or glass covering.

PROCEDURE

Follow all lab safety procedures for working with and disposing of a flame.

1. See FIG. 16-1.
2. Be sure the sand bags have been in the refrigerator for at least three hours.
3. Place the sand bags throughout the aquarium bottom.
4. Point the lamp into the aquarium.
5. Turn the lamp on.
6. Wait three minutes.

7. Place several small pieces of newspaper in a 500-milliliter beaker.
8. Light the paper. Wait several seconds, then blow out the flames.
9. Place the smoking paper in the aquarium. (If the aquarium is not smoky enough, repeat the above two steps.)
10. Cover the aquarium top completely with plastic wrap or a glass cover.
11. Watch what occurs. See FIG. 16-2.
12. Run water into the beaker with the smoldering newspaper to douse any flames or sparks.

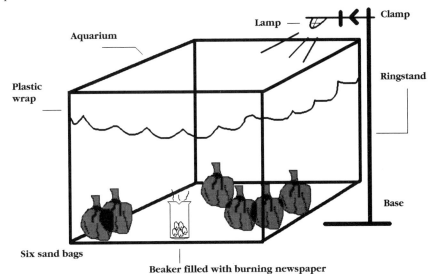

16-1 Lab setup

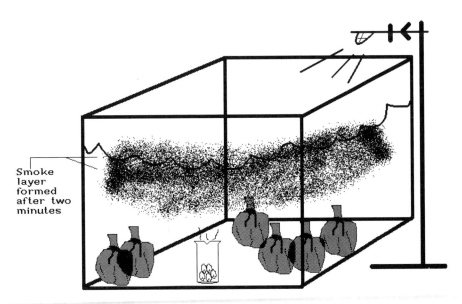

16-2 Example results

The temperature inversion 69

OBSERVATIONS

1. Where did the smoke layer eventually settle?
2. Why didn't the smoke layer move around in the aquarium?
3. What caused the inversion layer in this lab?
4. What was the purpose of the sand bags?

QUESTIONS/CONCLUSIONS

1. Why do inversion layers occur frequently in the San Fernando Valley or the Los Angeles Basin?
2. What conditions help to clear up a temperature inversion?
3. Where else do inversion layers occur?

POST-LAB ACTIVITY—LAB 16

Place several large beakers filled with crushed ice at both ends of an aquarium. In the middle, place a 500-milliliter beaker with near-boiling water. Seal off the aquarium from any exterior air currents. How quickly does the glass become fogged?

Use gloves while handling any heating devices. Now, take a 100-watt heating lamp and shine it on top of the "fog" in this aquarium. Does this light help to clear up the observed inversion layer? From what direction does the layer of condensation disappear (top to bottom or bottom to top)? How is the elevation of the inversion being affected by the heating lamp? Is the lamp's heat prolonging the layer's existence in the tank, or is it helping to clear it?

What clears a ground fog in the morning? Is the cooled surface of the earth partially responsible? Does the cooled surface promote the ground fog? Are the early morning sun's indirect rays striking the top of the layer capable of clearing this fog? How do ground fogs tend to dissipate (from ground to air or air to ground?)

Also see questions 1 and 55 in appendix C, "Environmentally Speaking."

17 Measuring the oxygen content of air

The chemical equation for the oxidation of iron is:

$$\text{Iron} + \text{Oxygen yields Iron Oxide}$$
$$Fe + O_2 \longrightarrow Fe_2O_3$$

1. Balance the equation.
2. How can the rusting of metal objects be prevented?

PURPOSE

The purpose of this lab is to approximate the oxygen content of air by rusting steel wool.

MATERIALS NEEDED

- One large test tube.
- Untreated (not oiled) steel wool.
- Pan.
- Water.
- Ringstand.
- Test tube clamp.
- Overhead projector marker.

PROCEDURE

1. See FIG. 17-1.
2. Measure the test tube length to the nearest 0.1 centimeter.
3. Record this length on FIG. 17-2.

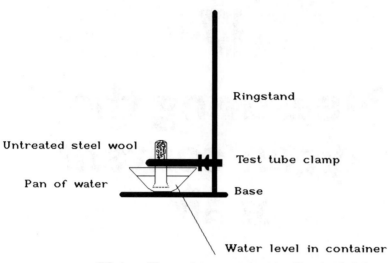

(Note: There is no water in the test tube, it is dry inside, and the air in the test tube is pushing the water away from its opening.)

17-1 Lab setup

1.	Length of test tube (base to lip) ....cm
2.	Height of water after three days ....cm
3.	Ratio of water height to test tube length (2 / 1) ....
4.	Percent of space filled with water (3) X 100 ....%
5.	Actual percentage of oxygen in air 20.946%

17-2 Data table

4. Be sure the test tube is completely dry.
5. Use tap water to slightly dampen a wad of steel wool.
6. Place the wad of steel wool inside the test tube.
7. Place the test tube upside down into the pan of water.
8. Clamp the tube to the ringstand.

72 The atmosphere

9. Leave the apparatus alone for three days.
10. After three days, use the projector marker to mark off the level of water inside the test tube. See FIG. 17-3.
11. Remove the test tube from the pan and the clamp.
12. Remove the steel wool.
13. Measure (to the nearest 0.1 centimeter) the distance from the opening of the test tube to the mark. (This is the height of the water.)
14. Record this height on FIG. 17-2.
15. Perform the calculations on FIG. 17-2.

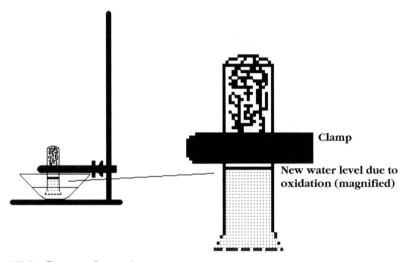

17-3 Closeup of test tube

OBSERVATIONS

1. Describe the appearance of the steel wool. What kind of change did it undergo: chemical or physical?
2. Why was the wad of steel wool wet with water?

QUESTIONS/CONCLUSIONS

1. How close is your value to the accepted value of 20.946 percent oxygen in air?
2. If the wool were left in the tube for a week, would the water level in the test tube become any higher? Why or why not?

POST-LAB ACTIVITY—LAB 17

Mass the dry steel wool to the nearest 0.01 gram, prior to placing it inside the test tube. Wet the wool slightly. After three days, dry the steel wool thoroughly (under a heat lamp or sunlight). Mass the wool again. What mass change do you expect, if any? How can this be explained?

 Also see question 60 in appendix C, "Environmentally Speaking".

18
Air exerts pressure

In this experiment, it would seem that the water should fall out of the glass. However, this is not so. Air in the atmosphere exerts a great deal of pressure—even in the classroom.

PURPOSE

The purpose of this lab is to demonstrate the air pressure pushing on a glass filled with water.

MATERIALS NEEDED

- Drinking glass.
- Water.
- 15-centimeter-by-15-centimeter cardboard or plastic square.

PROCEDURE

1. See FIG. 18-1.
2. Be sure the cardboard square is clean and dry.
3. Fill the drinking glass to the brim with water.
4. Place the card on the glass.
5. Perform the next step over a sink.
6. Quickly turn the glass directly upside down.
7. Watch what occurs.

OBSERVATIONS

1. Why isn't the water flowing out of the glass? (Hint: See FIG. 18-2.)
2. What is keeping the square securely on the glass?

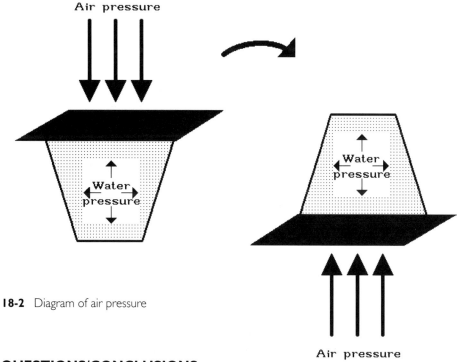

18-1 Lab setup

18-2 Diagram of air pressure

QUESTIONS/CONCLUSIONS

1. Define the term *air pressure*.
2. What happens to air pressure as altitude increases?
3. Would the results be any different if this lab were performed at 20,000 feet?
4. Why is the *atmosphere* in a jet passenger aircraft pressurized?

POST-LAB ACTIVITY—LAB 18

Will the cardboard or plastic square stay on no matter what the size of the opening of the jar is? Is there a limit to the size of the opening? Try different-sized jars and vary the amount of water in each. Use warm water, tap water, and ice-cold water to see if the temperature of the water affects the outcome.

How would the use of different liquids affect the results of this experiment? Try cooking oil. (*Caution! It can be messy/slippery.*) Also try salt water, sugar water, or any kind of carbonated soft drink.

Would a screen or handkerchief over the mouth of the glass work just as well as the cardboard/plastic square? Try it, but hold the screen or cloth tightly over the glass. Be sure to perform the preceding experiments over a tub or sink to catch any mishaps!

☆☆ ☆ FACTOIDS ☆ ☆

The following weather facts were paraphrased from the *1992 Encyclopaedia Britannica Book of the Year.*

- On July 1, 1992, the European Community prohibited the dumping of raw sewage at sea except in locations where it could cause "no harm."
- In 1991, a number of countries signed a protocol banning mining in Antarctica for 50 years.
- Initially felt to have caused global damage, the Kuwaiti oil fires (Persian Gulf War—1991) have caused severe local pollution but have caused no global consequences so far.
- Sulfur dioxide from the burning Kuwaiti oil fires made the rain in that area (North Arabian Gulf) more acidic than normal.
- The eruptions of Mt. Pinatubo in the Philippines generated enough dust and gas to occupy the upper atmosphere and hold back greenhouse warming for several years.
- The worst flooding in 30 years occurred in Austria in July/August, 1991.
- In 1991, more than a thousand toronados were reported in the United States. That number is well over the average of 748.
- Late in 1991, a deep ozone hole was again discovered over Antarctica, reaching the lowest record of ozone amount ever.
- Recent radar imagery from a U.S. space shuttle mission indicated the presence of ancient rivers beneath the Sahara in Egypt. Experimental drilling of this area showed enough fresh water (30 to 300 feet down) to support nearly a quarter-million acres of cultivation for the next 200 years.

Geophysical Features

Mountains play a very large part in local and regional weather. Oceans, deserts, and forests also have their role. Think how different the weather is in the Great Plains between the Rockies and the Appalachians.

The modification of wind is a profound example of how mountains play a role in weather. Wind cools as it travels up a mountain, and the moisture in the air condenses and falls as rain or snow. Thus, the windward side of a mountain is always lush and green; the opposite, lee side is often a desert. (The mountains near the Oregon coast are an example.) The next three labs are models of these orographic winds and the heating and cooling of land forms.

19

Heating & cooling of model land forms

Land forms heat up and cool down differently because of the angle of the sun, the season of the year, and what color each land form is. Darker "pieces"—dirt, desert, and rock—heat up faster than lighter-colored land forms—grassland and forests.

PURPOSE

The purpose of this lab is to observe the heating and cooling rates of models that represent different types of land forms.

MATERIALS NEEDED

- Eight plastic cups of different colors.
- Eight thermometers.
- A watch or clock with a second hand.
- Colored pencils.
- Sixteen white paper towels.
- Graph paper.
- Ruler.
- Ringstand.
- Lamp with 75-watt bulb and clamp.

PROCEDURE

1. See FIG. 19-1 and 19-2.
2. Figures 19-3 through 19-5 are tables of suggested colors to be used for this lab. If you use your own colors, see FIGS. 19-6 through 19-8.

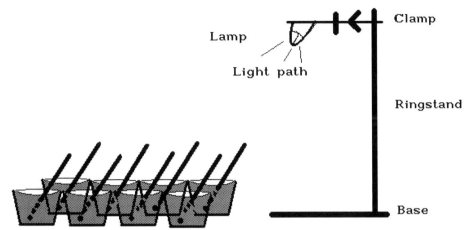

Eight cups with one thermometer and one crumpled paper towel in each cup.

19-1 Lab setup—lamp

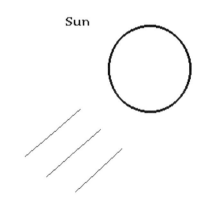

This laboratory exercise can also be performed out of doors in direct sunlight.

Eight cups with one thermometer and one crumpled paper towel in each cup.

19-2 Lab setup—sun

Heating & cooling of model land forms

COLOR OF CUP	BEGINNING TEMPERATURES (deg. C)
Orange	
Yellow	
Light Green	
White	
Blue	
Red	
Dark Green	
Black	

19-3 Suggested beginning temperatures

Heating temperature (deg. C)

COLOR OF CUP	1	2	3	4	5	6	7	8	9	10
Orange										
Yellow										
Light Green										
White										
Blue										
Red										
Dark Green										
Black										

For minutes 1 through 10

19-4 Suggested heating temperatures

Cooling temperature (deg. C)

COLOR OF CUP	11	12	13	14	15	16	17	18	19	20
Orange										
Yellow										
Light Green										
White										
Blue										
Red										
Dark Green										
Black										

For minutes 11 through 20

19-5 Suggested cooling temperatures

COLOR OF CUP	BEGINNING TEMPERATURES (deg. C)

19-6 Data table—beginning temperatures

Heating temperature (deg. C)

COLOR OF CUP	1	2	3	4	5	6	7	8	9	10

For minutes 1 through 10

19-7 Data table—heating temperatures

Cooling temperature (deg. C)

COLOR OF CUP	11	12	13	14	15	16	17	18	19	20

For minutes 11 through 20

19-8 Data table—cooling temperatures

Heating & cooling of model land forms

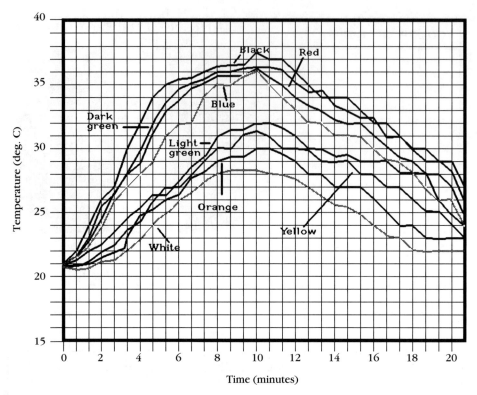

19-9 Example graph

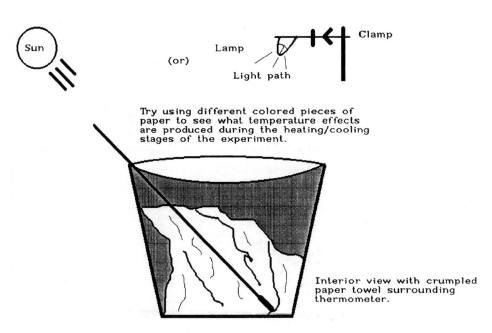

19-10 Suggestion for further exploration

82 Geophysical features

3. Place a thermometer into each cup.
4. Place one or two crumpled paper towels into each cup. Be careful of the thermometer.
5. Place the cups under a lamp or in direct sunlight.
6. On FIG. 19-3, record the beginning temperatures.
7. Turn on the lamp or place the cups in sunlight.
8. Start the clock.
9. At each minute, record the heating temperatures on FIG. 19-4.
10. At minute 10, record the temperatures and turn off the lamp.
11. If you're outside, bring all the cups back into the classroom.
12. On FIG. 19-5, record the cooling temperature each minute until minute 20.
13. See FIG. 19-9 for a sample graph of similarly recorded data.
14. Graph the recorded data on graph paper using appropriate scale.
15. Use a colored pencil corresponding to the cup color. For the white cup, use a regular pencil.

OBSERVATIONS

1. From your graph, which color cup heated the fastest? Which heated the slowest?
2. Did any of the cup temperatures plateau (level off) during the heating or cooling? Which one(s)?
3. Why were paper towels placed into each cup? Would the results of the experiment have been different if they had not been in each cup? Why or why not?

QUESTIONS/CONCLUSIONS

1. This lab was a model for different land forms heating and cooling at different rates. What possible land forms could these colors represent? (Name at least five.)

POST-LAB ACTIVITY—LAB 19

See FIG. 19-10 and try using different colored pieces of paper in each cup. Perform the experiment again with these colors. What sort of temperature trends do you expect? Did your graph of actual data differ from your predictions? How did the graph for this activity differ from the one you originally made for this experiment?

Also see questions 20 and 59 in appendix C, "Environmentally Speaking."

20
Heating & cooling of actual land forms

Similar to Lab 19, this lab shows the heating and cooling trends of actual samples. Since these samples are so small, visualize large portions of the earth with "parcels" of air being heated and cooled—ground, grasslands, oceans, lakes, and deserts. The microweather from such a regional area is quite difficult to forecast.

PURPOSE

The purpose of this lab is to observe the heating and cooling rates of samples of soil, grass, salt water, fresh water, and sand.

MATERIALS NEEDED

- Five large test tubes.
- Five single-hole rubber stoppers.
- Samples of soil, grass blades, salt water, fresh water (tap), and sand.
- Five thermometers.
- One 100-watt bulb with lamp stand.
- Test tube holder or ringstand with test tube clamps.
- A watch or clock with a second hand.
- Graph paper.
- Colored pencils.
- Ruler.

PROCEDURE

1. See FIG. 20-1.
2. Be sure to use glycerin or soapy water on the rubber stopper prior to placing it on the thermometer. See FIG. 20-2.

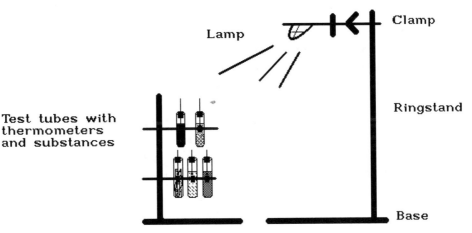

20-1 Lab setup

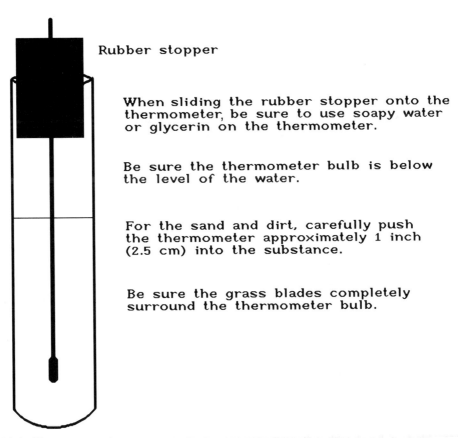

Rubber stopper

When sliding the rubber stopper onto the thermometer, be sure to use soapy water or glycerin on the thermometer.

Be sure the thermometer bulb is below the level of the water.

For the sand and dirt, carefully push the thermometer approximately 1 inch (2.5 cm) into the substance.

Be sure the grass blades completely surround the thermometer bulb.

20-2 Thermometer closeup

3. In each test tube, place one thermometer and one sample of the previously listed materials.
4. Place all five test tubes in clamps.
5. Secure the tubes/clamps under the lamp.
6. Record the beginning temperature on FIG. 20-3.
7. Turn on the lamp.
8. On FIG. 20-4, record the heating temperatures at each minute.
9. At minute 10, record the temperatures, then turn off the lamp.
10. On FIG. 20-5, record the cooling temperatures every minute until minute 20.
11. See FIG. 20-6 for a sample graph of similarly recorded data.
12. Graph recorded results on graph paper using appropriate scale.
13. Use a different colored pencil for each material.

Type of Material	Beginning temperatures (deg. C)
Soil	
Grass blades	
Salt water	
Fresh water	
Sand	

20-3 Data table—beginning temperatures

Heating temperature (deg. C)

Type of Material	1	2	3	4	5	6	7	8	9	10
Soil										
Grass blades										
Salt water										
Fresh water										
Sand										

For minutes 1 through 10

20-4 Heating temperatures

Type of material	Cooling temperature (deg. C)									
	11	12	13	14	15	16	17	18	19	20
Soil										
Grass blades										
Salt water										
Fresh water										
Sand										

For minutes 11 through 20

20-5 Cooling temperatures

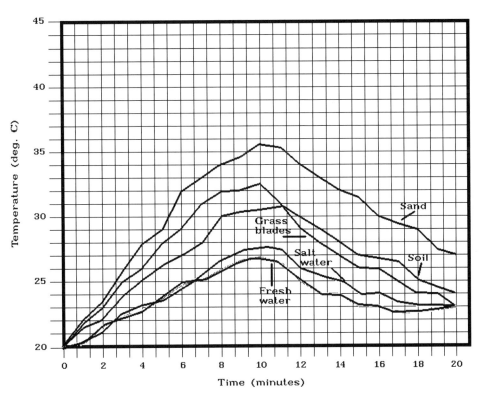

20-6 Example graph

Heating & cooling of actual land forms 87

OBSERVATIONS

1. Which material heated up the fastest? Which heated the slowest?
2. Which material cooled off the fastest? Which cooled the slowest?
3. Were there any differences in the heating and cooling of the saltwater and the freshwater temperatures?

QUESTIONS/CONCLUSIONS

1. At the beach on a sunny day, the sand is much hotter than the water. Why is this?

POST-LAB ACTIVITY—LAB 20

How different would the experiment data be if all of the solid materials used in this experiment were wetted slightly with water? What sort of heating/cooling curve would you predict?

Repeat the lab with new samples. Wet the solid portions until they are damp to the touch. Begin heating/cooling. Record and graph all data. How different is this graph from the original one?

Also see questions 20 and 59 in appendix C, "Environmentally Speaking."

21
How mountain ranges affect climate

When lifted over a mountain, moisture is deposited on one side of the mountain, while the other side is virtually dry. However, this model is quite limited; some moisture might end up on the lee side of the mountain due to the cooling of steam and not due to adiabatic cooling. In this experiment, the overall amount of precipitation will occur on the side with the beaker.

PURPOSE
The purpose of this lab is to observe moisture being lifted up by a model mountain.

MATERIALS NEEDED
- Aquarium.
- One 500-milliliter beaker.
- Two wood blocks.
- Calcium chloride (powder). *(Be sure to wear gloves and safety goggles when handling calcium chloride.)*
- Black construction paper.
- Cardboard.
- Plastic wrap or glass covering.
- Optional materials: mortar and pestle (to crush the calcium chloride grains into powder).

PROCEDURE

Caution: Observe all safety precautions while handling calcium chloride and other chemicals.

1. See FIG. 21-1.
2. Mount black construction paper on the cardboard.
3. Cut two cardboard panels until they fit into the width of the aquarium.
4. Tape both panels together.
5. Fill the 500-milliliter beaker with very hot water.
6. Place the cardboard panels into the aquarium. Brace the ends with the wood blocks.
7. The height of the panels should be just below the rim of the aquarium.
8. Put on gloves and safety goggles to handle the calcium chloride.
9. If the calcium chloride is granular, crush approximately 30 grams into powdered form.
10. Sprinkle powdered calcium chloride evenly over both panels. Cover them completely.
11. Place the beaker in the aquarium.
12. Cover the aquarium with plastic wrap or a glass cover.
13. Watch what occurs.
14. See FIG. 21-2 for example results.
15. Dispose of the calcium chloride by washing it down the sink with plenty of cold water.

OBSERVATIONS

1. Which side of the mountain showed moisture striking it first?
2. How quickly did the mountain become wet from the "rain?"
3. Which side had the most moisture deposited on it?

QUESTIONS/CONCLUSIONS

1. Define the term *orographic uplift* and describe the weather conditions of both sides of a mountain range that undergoes an orographic uplift.
2. Define the *lee side* of a mountain. What type of landscape is found there?
3. Define the *windward side* of a mountain. What type of landscape is found there?

POST-LAB ACTIVITY—LAB 21

How quickly would your model mountain become wet if a beaker of ice cubes were placed at the top? What if a beaker of ice cubes were placed at the bottom on the calcium chloride side, but a blow dryer (set on low-air only) was used? Would there be any difference in the time necessary to make the mountain wet? Try it!

Also see questions 11, 12, and 59 in appendix C, "Environmentally Speaking."

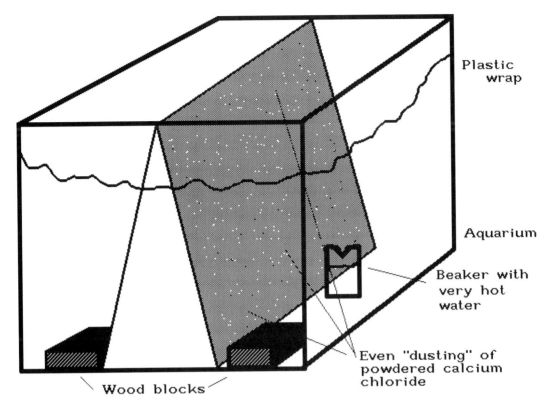

21-1 Lab setup

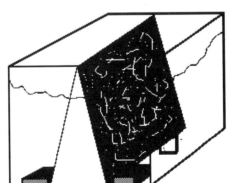

The calcium chloride will dissolve and become nearly transparent. This physical change is somewhat akin to the dissolving of powdered sugar.

Here, the calcium chloride has absorbed moisture from the beaker's water and shows signs of dampness.

21-2 Example results

How mountain ranges affect climate 91

PART 5

Weather Forecasting

Weather forecasting is both an art and a science. Art is involved in the day-to-day intuition of lay people sleuthing local weather for their own purposes. The science of weather forecasting involves a full spectrum of people—from the meteorologist's state-of-the-art radar/computer forecasting to the grade-school student's science fair project of measuring weather's big three: temperature, pressure, and humidity.

Yet, weather forecasting is now much more than recording those three facets. Indeed, as the following labs show, weather forecasting can be done from a variety of sources: the newspaper, radio, television, and you as well.

22
Long-term weather observations & graph

Ideally, this lab should be a 12-week exercise. A changing season and its accompanying weather is always exciting to measure and analyze. The graphs made here will give a climatic summary for your local area. See FIG. 22-1 for a six-day analysis of a front that brought rain to Los Angeles.

MATERIALS NEEDED
- Weather board (see Experiment 25) or an equivalent chart.
- Colored pencils (red, green, and blue).
- Graph paper.
- VHF weather radio (available inexpensively at local electronics stores).

PROCEDURE
1. See FIG. 22-2 for student-use chart.
2. See FIG. 22-3 for chart definitions.
3. Listen to the National Weather Service on 162.550 MHz, 162.525 MHz, 162.500 MHz, 162.475 MHz, 162.450 MHz, 162.425 MHz, or 162.400 MHz (24-hour VHF transmissions).
4. Record the current weather conditions in your area for 12 weeks (three months).
5. Set a consistent time each day to record conditions. (All of your science classes can use this data.)
6. See FIG. 22-4 for sky definitions.
7. See FIG. 22 5 for a sample graph of similarly recorded data.
8. Graph your results (using four weeks of recorded data per page of graph paper).

9. Using the appropriate scale, place humidity, pressure, and temperature on the vertical axis.
10. Use a blue pencil for humidity, a green pencil for pressure, and a red pencil for temperature.
11. Place the days on the horizontal axis.

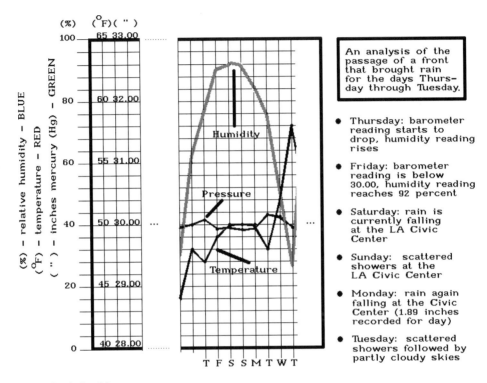

22-1 Analysis of front passage

OBSERVATIONS

1. What is the trend of the temperature during the first month?
2. According to your graph, what season appears to be approaching?
3. Does your graph support your answer to the above question? Why or why not?
4. What is the trend of the humidity during the three-month period?

QUESTIONS/CONCLUSIONS

1. Can you reasonably conclude that a drop in pressure and temperature means that precipitation will follow within one to two days? Does your data support this?
2. Can you identify any specific changes in the data just before a major change in the weather occurs?

Weather data for week of: ___/___/___ through ___/___/___

DAY	M /	T /	W /	T /	F /	S /	S /
TIME							
TEMPERATURE							
PRESSURE							
HUMIDITY							
WIND DIRECTION							
WIND SPEED							
SKY							
PRECIPITATION							

COMMENTS: (local-24 hour forecast; extended outlook, etc.)

22-2 Student-use chart

Weather data for week of: ___/___/___ through ___/___/___

DAY	M /	T /	W /	T /	F /	S /	S /
TIME	24-hour clock format – 7am = 0700, 1pm = 1300
TEMPERATURE	degrees Fahrenheit
PRESSURE	inches mercury (Hg)
HUMIDITY	percent (%), between 0 and 100
WIND DIRECTION	N E W S and NE SE NW SW
WIND SPEED	miles per hour (mph)
SKY	cloudy, partly sunny, partly cloudy, clear
PRECIPITATION	inches of rain or snow fallen: example – 1.02"

COMMENTS: (local 24-hour forecast; extended outlook, etc.)

22-3 Chart definitions

```
  cloudy - sky is completely covered with clouds, sun is obscured
partly sunny - more clouds than sun
partly cloudy - more sun than clouds
   clear - sky is clear, sun is not obscured
```

22-4 Sky definitions

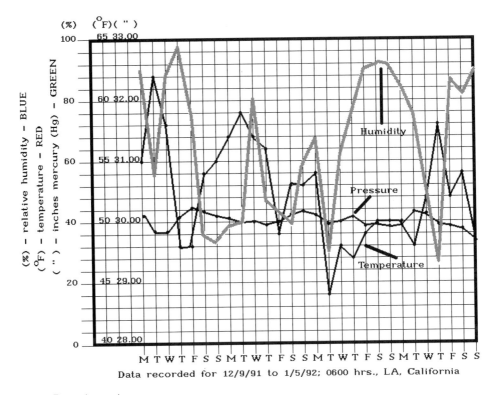

22-5 Example graph

POST-LAB ACTIVITY—LAB 22

Make your own analysis of the data obtained for your graph. Choose any week from the data (preferably a week in which precipitation occurred) and describe the trends of the temperature, pressure, and humidity. Indicate any precipitation types and amounts.

Location of readings: _____

Day/date: _____

_____

Long-term weather observations & graph 97

Day/date :_____
_____
_____

Day/date: _____
_____
_____

Day/date: _____
_____
_____

Day/date: _____
_____
_____

Day/date :_____
_____
_____

Day/date: _____
_____
_____

Also see questions 1, 7, 13 through 16, 18 through 23, 36, and 46 in appendix C, "Environmentally Speaking."

23
Satellite pictures from the newspaper

A newspaper photo of a cloud cover is certainly very useful in determining the weather over a large area. However, what is happening in your area right now (as compared to this larger region)?

PURPOSE
The purpose of this lab is to observe cloud/weather patterns from newsprint pictures.

MATERIALS NEEDED
- Newspaper with weather satellite photos.
- Glue stick or tape.
- Scissors.

PROCEDURE
1. See FIG. 23-1. Make copies as needed.
2. Use the daily newspaper map.
3. Cut the map out and tape or glue it to FIG. 23-1.
4. Record the current weather conditions on FIG. 23-1.
5. Keep a log for two weeks to observe changing weather patterns.
6. These maps could be placed on a plastic transparency for an overhead projector and used for the entire class.

```
┌─────────────────────────────────────────────────────────────────┐
│  DAY OF PICTURE: _____            │
│                                                                 │
│  DATE OF PICTURE: _____            │
│                                                                 │
│  NAME OF NEWSPAPER: _____            │
│                                                                 │
│  REGION OBSERVED (FROM NEWSPAPER PICTURE): _____           │
│                                                                 │
│  YOUR OUTSIDE WEATHER CONDITIONS (circle all that apply):       │
│    tornado    thunderstorm    hurricane    tropical storm   flood │
│           hail         sleet        snow         rain            │
│    fog      smog     smoke     haze     drizzle    marine layer  │
│      overcast    cloudy    partly sunny    partly cloudy   clear │
│        windy         hot         cold         warm        cool   │
└─────────────────────────────────────────────────────────────────┘
```

(Affix satellite photo here.)

23-1 Satellite photo page

OBSERVATIONS

1. How far does a front (cold or warm) appear to move each day?
2. What direction do the fronts travel to and from, according to the satellite photos?

QUESTIONS/CONCLUSIONS

1. Is it possible for you to predict the weather conditions for your area based on the movement of the fronts observed in your pictures?
2. In what direction do fronts travel in the Southern Hemisphere?
3. Can you determine in how many days the front you're observing will move into another region? (Hint: use the front's velocity (speed and direction).)

POST-LAB ACTIVITY—LAB 23

Obtain newspaper satellite photo maps for one week before and one week after Hurricanes Andrew (August, 1992) and Iniki (Sept., 1992). Also obtain newspaper clippings to create an eyewitness journal.

On a separate sheet of paper, answer the following questions:

1. Are the patterns for hurricanes different in different regions?
2. What are the latitudes/longitudes of Hawaii and Florida?
3. What time of day did each hurricane strike their respective states?
4. How fast was each storm traveling?

5. What was the highest wind gust for each? What was the average strongest wind?
6. How were these hurricanes similar to each other?
7. Do hurricanes striking in the same latitudinal area follow the same path?
8. Do hurricanes follow the same direction in the Northern Hemisphere?
9. Is there a difference in the two hurricanes mentioned above, as far as intensity (wind speed/direction?)
10. How is damage related to the path taken by a hurricane?

Also see questions 1 through 3, 13 through 16, 18, 19, 29, 36, and 46 in appendix C, "Environmentally Speaking."

☆ ☆ FACTOIDS ☆ ☆

The following facts were paraphrased from newspapers and television:

- Hurricane Andrew, "the storm of the century," was the most expensive storm ever in U.S. history.
- Every acre of land on this planet has at least one molecule of dust from every other acre of land.

☆ ☆ ☆ ☆ ☆ ☆ ☆ ☆ ☆ ☆ ☆ ☆ ☆ ☆ ☆ ☆ ☆

24
"Spot" 24-hour weather predictions

Local weather stands alone as perhaps the most difficult forecasting of all. It's one thing to give a quality three-to-five-day forecast coast-to-coast. However, it's the short-range weather covering the next 18 to 48 hours that influences the life of the nation in all its phases and is more difficult to predict. Why? Because of the numerous minute breezes and temperature, pressure, and humidity changes that are so peculiar to a given locale. The charts in this lab generalize for the United States, so use them in addition to your common sense.

PURPOSE

The purpose of this lab is to use existing NWS (National Weather Service) data to make your own forecast.

MATERIALS NEEDED

- VHF radio with weather frequencies (National Weather Service—162.550 MHz, 162.525 MHz, 162.500 MHz, 162.475 MHz, 162.450 MHz, 162.425 MHz, 162.400 MHz.)

PROCEDURE

1. See FIGS. 24-1 and 24-2.
2. See FIG. 24-3 for descriptions of cloud types.
3. Copy the current weather conditions from a local TV news broadcast or from the National Weather Service.
4. Using wind direction, barometric pressure, and the direction the barometer is taking, make a forecast.

24-1 Chart A

Barometer reading	WIND DIRECTION							
	N	NE	E	SE	S	SW	W	NW
Above 30.20								
RR/RS	fair, cooler	fair	fair, cooler	fair	fair	fair	fair	fair, cooler
Steady	N/C	N/C	N/C	N/C	N/C	N/C	N/C	N/C
30.00 to 30.20								
RR/RS	fair, cooler	fair	fair, cooler	fair, warmer	fair	fair	fair, cooler	fair, cooler
Steady	N/C	N/C	N/C	N/C	N/C	N/C	fair	fair
Below 30.00								
RR/RS	fair, cooler	fair, cooler	fair	fair	fair	fair	fair, colder	fair, colder
Steady	N/C	N/C	N/C	N/C	N/C	N/C	N/C	N/C

RR – rising rapidly N/C – no change FR – falling rapidly
RS – rising slowly FS – falling slowly

24-2 Chart B

Barometer reading	WIND DIRECTION							
	N	NE	E	SE	S	SW	W	NW
Above 30.20								
FR	cloudy, rain	rain, wind	rain, wind	cloudy, rain	rain	cloudy, warmer	cloudy	rain, wind
FS	cloudy	wind, or rain	wind, rain	cloudy, warm	cloudy	fair, warmer	fair	fair
30.00 to 30.20								
FR	rain, colder	rain, wind	rain, wind	rain, warm	rain, wind	rain, colder	rain	cloudy
FS	cloudy	rain	rain	rain, warm	rain, warm	cloudy, warm	fair, warm	unsettled
Below 30.00								
FR	danger, gale!	danger, gale!	severe storm	severe storm	storm, cold	rain	rain	unsettled
FS	unsettled	rain, cooler	rain	rain	rain	unsettled	unsettled	unsettled

RR – rising rapidly N/C – no change FR – falling rapidly
RS – rising slowly FS – falling slowly

OBSERVATIONS

1. Did your forecasts match the actual weather?

QUESTIONS/CONCLUSIONS

1. What other information does a meteorologist require to make an accurate forecast?
2. How have computers and satellites aided in the forecasting of weather?

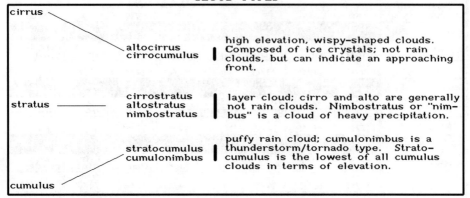

24-3 Description of cloud types

POST-LAB ACTIVITY—LAB 24

Maintain a log of your "spot" predictions and the actual outcome. Keep the log for at least four weeks. Calculate your percentage of accuracy:

$$\frac{\text{number of times correct}}{\text{number of days recorded}} \times 100 = \text{percent (\%) accuracy}$$

This percentage will be quite approximate because your only access to data will be that which is provided by the NWS, and your forecasts are made without the far-ranging equipment that's at the heart of the NWS.

Also see questions 1 through 3, 16, and 36 in appendix C, "Environmentally Speaking."

25
The weather data station

This project can be used in conjunction with Lab 22. The data station makes an excellent room-sized display board that is highly visible throughout the classroom.

PURPOSE
The purpose of this lab is to construct and use a weather data system for long-term weather observations.

MATERIALS NEEDED
- One 3-foot-by-5-foot cork backboard with frame.
- Erasable markers (different colors).
- Plastic-coated cardboard (or a 3-foot-by-5-foot "white board").
- Half-inch black plastic letters/numbers (self-adhesive Roman or Gothic).
- Black permanent-marker pen (broad tip).
- T square.
- Paper cutter.
- White glue.

PROCEDURE
1. See FIG. 25-1.
2. Use the T square and the permanent marker to draw division lines on the plastic-coated cardboard. If the "white" board is used, draw division lines on its surface.
3. Place the letters/numbers on the board.
4. With the paper cutter, cut the appropriate size of coated cardboard to fit the backboard.
5. Glue the cardboard to the cork backboard.

Weather data for week of: __/__/__ through __/__/__							
DAY	M	T	W	T	F	S	S
TIME							
TEMPERATURE							
PRESSURE							
HUMIDITY							
WIND DIR.							
WIND SPEED							
SKY							
PRECIPITATION							

COMMENTS: (local 24 hour forecast; extended outlook, etc.)

Optional items:

Digital clock

VHF weather radio

Regional map

These items can be attached to the board using velcro tabs.

If possible, laminate the regional map so erasable markers can be used on the plastic surface.

25-1 Weather data model

6. Mount the backboard in the room or in a window for class observation.

7. See Lab 22 to begin acquiring data and analyzing it.

The following are optional:

1. Glue a map of your state/city in an appropriate area on the board.
2. Obtain a weather radio (available inexpensively from specialty electronics stores.) With velcro strips, affix the radio to the board.
3. Other items: small, inexpensive, battery-operated digital clock, and photographs of the three basic cloud types (cirrus, stratus, and cumulus.)

POST-LAB ACTIVITY—LAB 25

As an additional long-term assignment for a group of students, have them take pictures of at least ten cloud types. Be sure the pictures are at least 4 inches by 6 inches (for viewing by a large group of people.)

As a rule of thumb, keep in mind these cloud types:

- Cirrus, cirrostratus, cirrocumulus.
- Altocumulus, altostratus,
- Stratocumulus, stratus, nimbostratus.
- Cumulus, cumulonimbus.

You can find detailed descriptions of each of these types of clouds in a number of meteorology texts.

Also see question 1 in appendix C, "Environmentally Speaking."

Appendix **A**

Tracking storms of the season

This appendix contains suggestions for long-term activities for middle-school and high-school students in the following subject areas: math, science, economics, and government. Many schools are now year-round, and any of these projects would be quite excellent to perform as an individual project through any seasonal period. The subjects of the chapter are as follows:

- Hurricane/tornado tracking. The student records a national view of these storms from commercial and amateur radio, TV, and newspapers/magazines. Clippings of the written articles describe paths, damage, casualties, and impact on the region's economy and law and order. *(Never attempt to follow or witness the weather phenomena firsthand.)*
- Thunderstorm watch. The student analyzes local and regional weather and, from the various media sources, records data on all thunderstorms in the local area. Clippings of synoptic maps and related articles are maintained by the student to create a journal of the thunderstorm season. *(Keep in mind that thunderstorms can be dangerous and it's not advisable to be outside in a thunderstorm. However, thunderstorms can be safely viewed from cars if the winds are relatively low (60 miles per hour or less). The two safest places to be during thunderstorms are in houses and cars.)*
- Snowstorm season. In areas with snowfall, the student records, graphs, and analyzes monthly snowfall totals. All storms are tracked from the beginning to the end of each season. *(Snowstorms also pose risks, and it's not advisable to stay outside during extreme snowstorms.)*
- Rainy season. In areas with rainfall, the student records, graphs, and analyzes weekly and monthly rainfall totals.
- Pollution index for air quality. Although not a storm-related subject, this factor still impacts the environment. On a daily basis, the student records data from media sources counting the parts per million (ppm) of airborne particulates. The student then graphs and analyzes the data.

HURRICANE/TORNADO TRACKING

This can be performed by an individual or a group of up to four students. The goal is to trace, maintain, and update data from these violent storms through a given period of time (this will be set by the instructor).

Obtain a map of the United States from a local automobile club. Laminate the map in clear plastic. Erasable markers of various colors, and adhesive multicolored stars (⅜ inch) will work quite well in labeling each storm.

Make extra copies of Labs 22, 23, and TABLES A-1 and A-2 to obtain all data/remarks. Watch TV, listen to the radio (including information received from any amateur radio transmission), and read the newspapers to obtain any information regarding any event (weather, social, or political) that is related to the storm being watched.

Table A-1 Hurricane tracking worksheet

For the month of: _____,199____

Data tracked from: _____ to: _____
Information received from: _____
                              (give source)
Name of hurricane: _____
Location where spotted first: _____
Date/location where landfall occurred:_____
Date/location hurricane was downgraded to tropical storm:
_____
Did tornados result from this hurricane? (If yes, see worksheet "tornado tracking" and continue. If no, continue.)
Estimated damage (dollar amount and size of stricken area):
_____
Casualties: _____
How many days before stricken area was declared a disaster area by the federal government?
_____
Date federal troops/American Red Cross were called in for relief effort assistance:
_____
(Attach any news clippings to this page or on a separate sheet.)

Table A-2 Tornado tracking worksheet

For the month of: _____,199____

Data tracked from: _____ to: _____
Information received from: _____
                              (give source)

Table A-2 Continued.

Was/Were this/these tornado(s) spawned by a hurricane or thunderstorm? (If yes, attach this sheet to the worksheet "hurricane tracking" or "thunderstorm watch" and continue. If no, continue.)

Location where spotted first: _____
Length of time vortex was on ground: _____
Width of swath:_____
Estimated wind intensity (mph or km/h): _____
Estimated damage (dollar amount and size of stricken area):
_____
Casualties: _____
How many days before stricken area was declared a disaster area by the federal government?
_____
Date federal troops/American Red Cross were called in for relief effort assistance:
_____
(Attach any news clippings to this page or on a separate sheet.)

Keep in mind that data obtained during the weather event must be preceded by and followed by information from prior to and after the storm. This way, the view of such an event is complete and gives a truer picture of the events leading up to and following the storm. A rule of thumb is to obtain articles/clippings one week before and one to twelve weeks after. However, the aftermath of any violent storm can certainly last for months (or even years). This will be quite evident as citizens rebuild Florida or Kauai, Hawaii from the devastating Hurricanes Andrew and Iniki.

THUNDERSTORM WATCH

As with the hurricane/tornado tracking, this activity can also be performed by one to four students. Copy Labs 22, 23, 24, and TABLE A-3 to use as worksheets. Be sure to list sources of any information: National Weather Service, TV, radio (amateur as well as commercial) and newspaper.

Table A-3 Thunderstorm watch worksheet

For the month of: _____,199_____

Data tracked from: _____to: _____
Information received from: _____
                              (give source)

Table A-3   Continued

Time of thunderstorm: _____ to: _____

Did tornados result form this thunderstorm? (If yes, see worksheet "tornado tracking" and continue. If no, continue.)

Did flash floods result from this thunderstorm? (If yes, describe locations and intensity on a separate sheet of paper or attach any articles.)

Casualties: _____

(Attach any news clippings to this page or on a separate sheet.)

SNOWSTORM SEASON

The season can be broken up into months/weeks to be assigned to blocks of students. Copy Labs 22, 23, and 24 for worksheets. Make graphs of all snow precipitation amounts on a daily/weekly/monthly basis. Ideally, you should use three to six months (or the official season dates).

RAINY SEASON

The season can be broken up into months/weeks (similar to the snowstorm season) and assigned to various students. Copy Labs 9, 22, 23, and 24 for worksheets. Make any graphs showing amounts fallen. Use the snowstorm season as a guide.

POLLUTION INDEX FOR AIR QUALITY

Any number of months from a particular season can be observed by a set number of students. A daily recording of any pollution level (of particulates) in parts per million (ppm) is given out over TV, radio, or newspaper by an agency responsible for such information. In California, the agency is the Air Quality Management District (AQMD).

Copy Labs 22, 23 and 24. In the comments section of Lab 22, be sure to describe the pollution level. Graph all data (ppm on y-axis, and days on the x-axis). Keep this in a journal form. Is there any link to pollution levels and politics (especially during an election year)?

Appendix B
Weather trends

The following worksheets will give math/science students an opportunity to graphically analyze their own local/regional weather data. Individually graphed trends of temperature, barometric pressure, and relative humidity, as well as amounts of precipitation will enable students to see a link between these trends and weather forecasting. The crucial skills of correctly graphing data are also reinforced.

Use FIG. 22-2 (student-use chart) for obtaining data for all of the following worksheets. Colored pencils are a plus for making a graphed line or bar "stand out." Make extra copies of FIG. 22-2—it shows only weekly recorded amounts. Write the day/date in the indicated available spaces.

It's important to make all data recordings at a consistent time each day. Be sure to make sky observations of your local weather. To avoid frustration in possible inconsistent readings, adhere to either the NWS (National Weather Service) station giving the data, or to a local radio/TV broadcast. Consistency is the key to successful observations and analyses of them.

Connect all plotted points with a smooth, dark line. If shading is required, shade smoothly and evenly. For the temperature worksheets, choose a wide enough range to include all your temperatures for the observed month. Use the indicated scale to write in your locally observed temperatures. Keep the scale consistent throughout the weeks of data. See FIGS. B-1, B-2, and B-3.

For the barometric pressure worksheets, connect the plotted points with a smooth line. Write a one-word description of the weather for that day (fair, sunny, cloudy, rain, snow, hail, etc.) See FIGS. B-4, B-5, and B-6. For the pressure/wind direction graph, connect the plotted points of pressure with a smooth line. Color in the appropriate wind direction "petal." See FIGS B-7, B-8, and B-9. For the relative humidity bar graph, shade in the bar to the correct relative humidity reading with a colored pencil. See FIGS. B-10, B-11, and B-12. For the wind direction/wind speed graph, use the indicated scale for local observed wind directions and speeds. See FIGS. B-13, B-14, and B 15. For the weather statistics chart, write in the dates to follow the calendar month you're analyzing. Make all percent calculations rounded to nearest percent. See FIGS. B-16 and B-17.

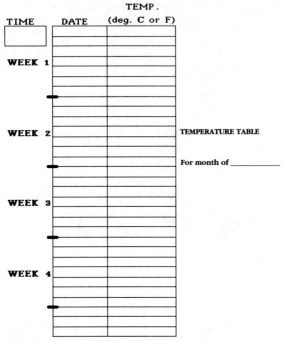

B-1 Temperature table

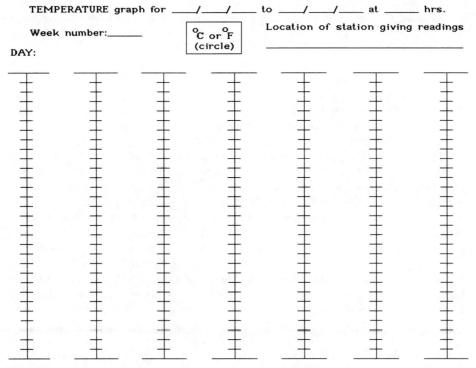

B-2 Temperature graph

112 Appendix B

For the precipitation amounts graph, use the indicated scale for rain in inches, snow in feet/inches. Total the amount of rain recorded. Total the amount of snow recorded. Connect only the days with rain with a straight line. Connect only the days with snow with a straight line. Don't cross over into the snow or rain portion if the phase of the precipitation changes from the previous day. Instead, draw a red dotted line between days of phase change and write "PHASE CHANGE." See FIGS. B-18, B-19, and B-20.

With all graphs, now it's possible to "read" and correlate the data with what actually happened. For instance:

- What was the local weather when barometric pressure dropped with an easterly wind?
- When pressure rose, followed by a westerly wind, what weather conditions were observed?
- When humidity fell/rose to less than/more than 30%/90%, what were the weather conditions?
- In fair/stormy weather, what direction do winds mostly tend to come from?

There are a great deal of other logical, numerical connections between your data and the observed weather conditions. A strong skill in interpreting these graphs will lead to an equally strong weather forecasting skill. Enjoy.

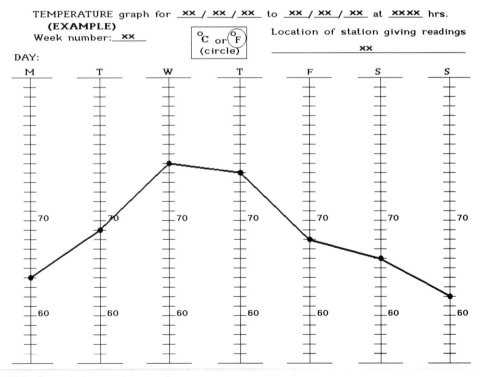

B-3 Example temperature graph

Weather trends 113

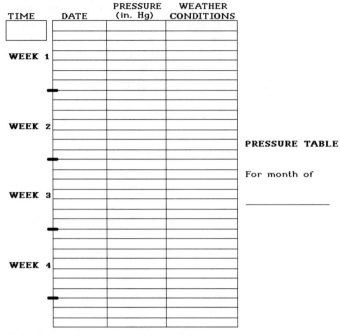

B-4 Pressure/weather-conditions table

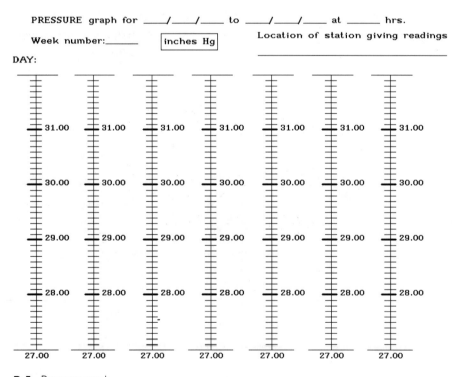

B-5 Pressure graph

114 Appendix B

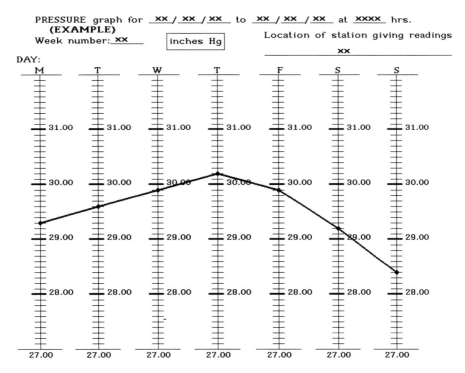

B-6 Example pressure graph

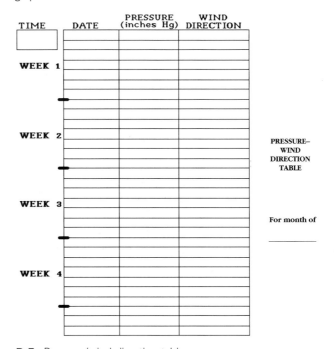

B-7 Pressure/wind-direction table

Weather trends 115

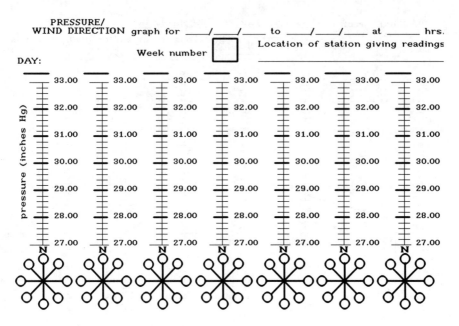

B-8 Pressure/wind-direction graph

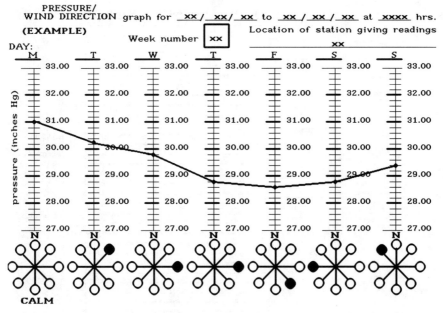

B-9 Example pressure/wind-direction graph

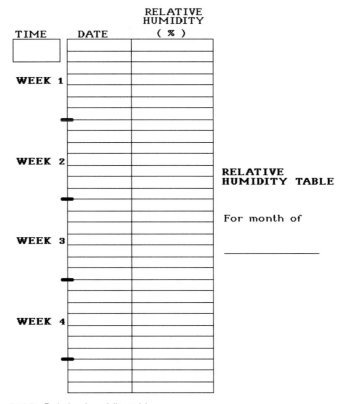

B-10 Relative humidity table

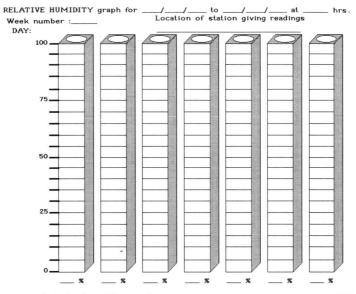

B-11 Relative humidity graph

Weather trends 117

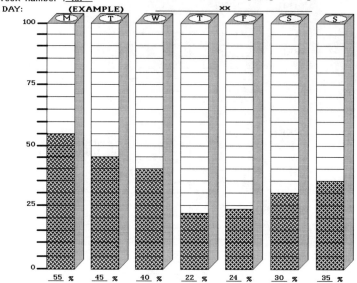

B-12 Example relative humidity graph

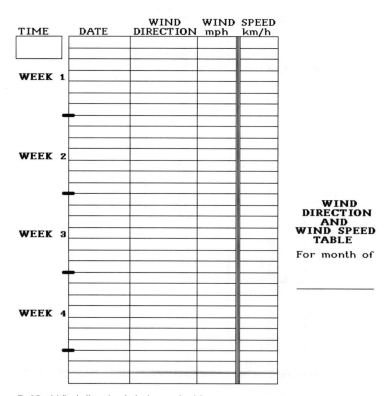

B-13 Wind direction/wind-speed table

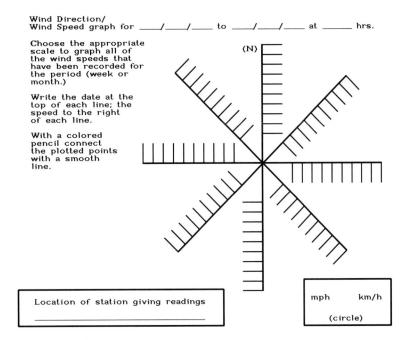

B-14 Wind direction/wind-speed graph

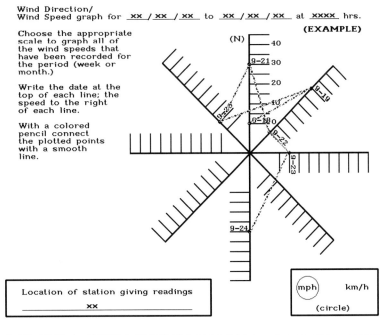

B-15 Example wind direction/wind-speed graph

Weather trends 119

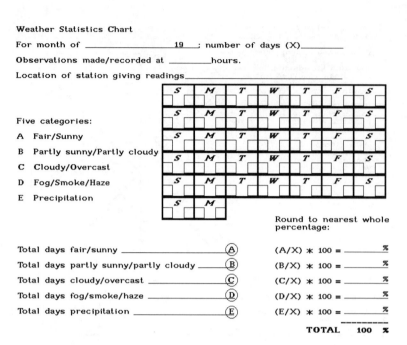

B-16 Weather statistics chart

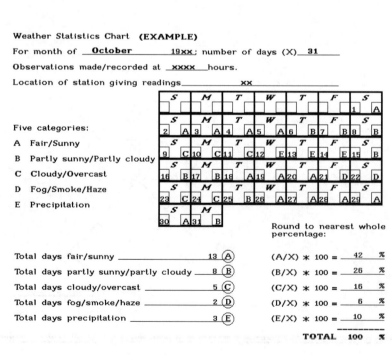

B-17 Example weather statistics chart

TIME	DATE	PRECIPITATION AMOUNTS					
		Rain	in.	Snow	ft.	in.	
		Rain	in.	Snow	ft.	in.	
		Rain	in.	Snow	ft.	in.	
WEEK 1		Rain	in.	Snow	ft.	in.	
		Rain	in.	Snow	ft.	in.	
		Rain	in.	Snow	ft.	in.	
		Rain	in.	Snow	ft.	in.	
		Rain	in.	Snow	ft.	in.	
		Rain	in.	Snow	ft.	in.	
		Rain	in.	Snow	ft.	in.	
WEEK 2		Rain	in.	Snow	ft.	in.	**PRECIPITATION AMOUNTS TABLE**
		Rain	in.	Snow	ft.	in.	
		Rain	in.	Snow	ft.	in.	
		Rain	in.	Snow	ft.	in.	For month of
		Rain	in.	Snow	ft.	in.	
		Rain	in.	Snow	ft.	in.	
WEEK 3		Rain	in.	Snow	ft.	in.	_____
		Rain	in.	Snow	ft.	in.	
		Rain	in.	Snow	ft.	in.	
		Rain	in.	Snow	ft.	in.	
		Rain	in.	Snow	ft.	in.	
		Rain	in.	Snow	ft.	in.	
		Rain	in.	Snow	ft.	in.	
WEEK 4		Rain	in.	Snow	ft.	in.	
		Rain	in.	Snow	ft.	in.	
		Rain	in.	Snow	ft.	in.	
		Rain	in.	Snow	ft.	in.	
		Rain	in.	Snow	ft.	in.	
		Rain	in.	Snow	ft.	in.	
		Rain	in.	Snow	ft.	in.	

B-18 Precipitation amounts table

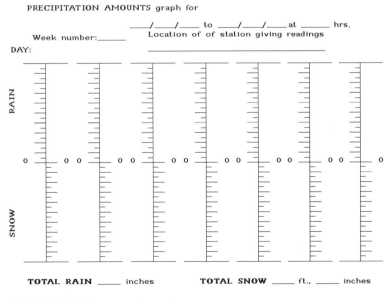

D-19 Precipitation amounts graph

Weather trends

PRECIPITATION AMOUNTS graph for
(EXAMPLE)   xx / xx / xx to xx / xx / xx at xxxx hrs.

Week number: xx   Location of of station giving readings

TOTAL RAIN __3__ inches   TOTAL SNOW __2__ ft., __0__ inches

B-20 Example precipitation amounts graph

Appendix C
Environmentally speaking

The following questions are applicable to more than one laboratory exercise. The questions combine several ideas and relate them to aspects of the earth's environment and ecology. While there is no "one good answer" for each, your students (and you) will find these questions quite challenging. A student might be curious enough to pursue a science fair project based on researching the answer to one (or more) of these questions. There might be questions where there are no related labs from this book. If that's the case, begin your research in a high-school or university library.

Suggested publications to begin research include: *National Geographic, Weatherwise, Discover, PROBE!, TIME, Newsweek, Scientific American, OMNI, Science News, The Science Teacher*, and *Psychology Today*. Other magazines that peripherally relate to weather are: *Field and Stream, Popular Science, Leisure and Travel, Car and Driver, Popular Photography, Consumer's Digest, Nation's Business, Weekly World News, Smithsonian, Homeowner,* and *Business Week*. Also, don't forget to check your local library for encyclopedias, the *Reader's Guide to Periodical Literature,* and *Reader's Guide Abstracts*.

1. How has Mt. Pinatubo (in the Phillipine Islands) contributed to the El Niño effect since 1991? Related labs: 3, 5, 6, 13, 16, 22, 23, 24, 25.
2. Have the Kuwaiti oil fires significantly changed the weather for agricultural activities in the Arabian Desert? Related labs: 23, 24.
3. If another eruption of Mt. St. Helens occurs, how can this possibly affect the logging industry for the Pacific Northwest? Related labs: 23, 24.
4. Are there other locations throughout the world where the ozone layer is showing signs of "thinning?" (Antarctica is one location.) Related reading: *TIME, National Geographic*.

5. Has the pollution from motor vehicles caused any ozone destruction? Related reading: *TIME, Science News, National Geographic.*
6. How are CFCs (chlorofluorocarbons) being reduced in your city? Related reading: *Discover, Science News.*
7. Why is the amount of yearly rainfall in Southern California dropping? Could it possibly be related to sunspot activity? Is there any possible relation to meteor shower activity? Related labs: 22.
8. Is the pH (acid/base) content of rain different from storms that strike your region from different directions? Related reading: *Weatherwise.*
9. As rain forests in the South American continent are being depleted, how is this biome being transformed? Will its depletion result in a desert area? Related lab: 3.
10. Are rainmaking efforts somewhat successful? Are clouds being seeded too much or too little? Is there a political influence over when to seed or not seed? Related reading: *PROBE!, Discover.*
11. Silver iodide has been used for the "seed" in rainmaking. However, it is a corrosive chemical. Develop an experiment that uses the dust of tea leaves as the condensation nuclei for ice crystal formation. Related labs: 21.
12. Where is cloud seeding most successful? Can it be used, say, in the Arabian Gulf to influence agriculture? At what cost (i.e., economically, politically)? Is it possible that rain in one area might create an irreversible weather condition in another area (extreme drought where there was none before)? Related labs: 21.
13. How will NEXRAD (National Weather Service next generation radar) be helpful in forecasting microbursts for airport traffic? Related labs: 22, 23.
14. Hurricane Andrew (August 1992) has been the United States' costliest natural disaster in United States history. Why are storms such as Andrew (others: Camille and Hugo) so violent? What has been the trend of these violent storms in the past three decades? How could the damage costs have been reduced? Related labs: 22, 23.
15. What is happening to the United States' coastline? Can anything be done to slow down this gradual erosion of beachfront property? Related labs: 22, 23.
16. Hurricane Iniki (September 1992) has been called Hawaii's worst storm this century. How could the citizens have better prepared themselves for a hurricane's landfall? Related labs: 22, 23, 24.
17. How has weather entered the political arena? (i.e., The Earth Summit was attended by a number of representatives in Brazil in June of 1992.) What was the outcome of the meetings? Related reading: *TIME, Science News.*
18. Has weather modification been used as a weapon against other countries? Related labs: 22, 23.
19. Is it ethical for the government to manipulate the weather for gain? Related labs: 22, 23.
20. What is the trend toward the land growth of desert regions? Are they increasing in size? Is human activity affecting these areas? Related labs: 19, 20, 22.

21. Are famine and weather cycles related? What countries show the greatest chances of such famines occurring? Should weather be modified to minimize such catastrophes? Related labs: 22.
22. How has food production in the United States been affected by weather changes? Related labs: 22.
23. How has the change in farming methods affected the weather due to the loss of topsoil on farmland in the United States? Related labs: 22.
24. How have designs in the clothing manufacturing industry been changed in the use of innovative materials for use in certain regions (i.e., water repellent clothing)? Related reading: *National Geographic*.
25. How does weather affect regional population locations across the United States? Related reading: *TIME*.
26. Is our industrial revolution changing weather patterns in the United States? As one example, trace the history of the railroad in the United States since 1869. (Transcontinental linkage occurred in this year at Promontory Point, Utah.) Related reading: *Discover, Science News*.
27. How has weather affected a household's waste (liquid and solid) output? Related reading: *OMNI, Discover*.
28. How have wars been decided by weather influences? (D-Day, Napoleon's invasion of Russia, and Hitler's invasion of Russia are some examples.) Related reading: *National Geographic*.
29. Is there any "weather" on the moon? Where could there be any possible "pockets" of water on the moon? Related labs: 23.
30. How has recycling of waste materials in the United States (glass, paper, plastics) been influenced by the weather? Related reading: *OMNI, Discover*.
31. What food items are purchased seasonally? How does weather affect the availability of these items? Related reading: *Discover, Science News*.
32. Are religious ceremonies linked to the weather? In what cultures is any link still practiced? Related reading: *National Geographic*.
33. How has weather affected people's beliefs and superstitions of weather forecasting? Related reading: *National Geographic*.
34. In what possible ways did the weather influence human discovery of fire? Related reading: *National Geographic*.
35. What's nature's way of "fixing" the earth? (Every few years, lightning-induced fires permit certain tree seedlings to reproduce.) What other events occur? Is it appropriate for humans to "tamper" with these occurrences? Related reading: *Weatherwise, Discover*.
36. How does weather affect leisure time in the United States? When does the greatest amount of auto/train/plane travel occur? How does this affect the economy of a region? If weather is not favorable in a certain region (Hurricane Andrew striking Florida or Hawaii's Hurricane Iniki) how does this affect that state's economy? Related labs: 22, 23, 24.
37. How has weather affected language growth and development in certain countries? Related reading: *National Geographic*.

38. What geophysical features of the United States affected the growth, movement, and expansion of the United States population in the last 100 years? Related reading: *National Geographic, Weatherwise.*

39. How has weather influenced the discovery of new materials, food production, and medicines? Related reading: *Science News, OMNI, Discover.*

40. How has weather affected the designing of safer and more durable automobiles? Trace the safety features and use of modern materials of the automobile from 75 years ago to the present day. The headlight system and the rubber tire are two examples. Related reading: *Science News, TIME.*

41. Are buildings (skyscrapers) more "environment friendly" today? That is, do they impact less on the environment by reflecting less light, producing less waste (gas or liquid or solid?) Related reading: *TIME, Discover.*

42. How has the proliferation of "mini-malls" contributed to weather changes? Where a vacant lot once stood, a collection of stores producing solid, liquid, and gaseous waste now influence the microweather of small regions. In what ways do these buildings affect the regional weather? Related reading: *TIME, Discover.*

43. How has the study of medicine been influenced by the weather? Related reading: *National Geographic.*

44. How has weather affected the entertainment industry? Related reading: *TIME, Weatherwise.*

45. How is/was the harvest season influenced by weather? (Trace it from 100 years ago to today, i.e., the use of the cotton gin, mechanical threshers, invasion of locusts denuding crops, and the heartbreak of replanting those crops.) Related reading: *Discover, National Geographic.*

46. How has the weather affected contributions to charitable organizations? Related labs: 22, 23.

47. How has education been influenced by the weather? Related reading: *The Science Teacher, Discover, OMNI, TIME, Science News.*

48. How have the notions of the "family" and family values been influenced by the weather? Related reading: *TIME, Discover, National Geographic.*

49. How does the weather influence the psychological makeup of an individual? Do different living conditions in different regions influence human emotional growth? Related reading: *Psychology Today, National Geographic.*

50. How has weather affected the transfer of utilities in your region to your home? Are electrical lines above the ground? Why or why not? What influences the location and positioning of natural gas or water lines for your usage? Related reading: *OMNI, TIME.*

51. How has weather influenced the manufacture of children's toys? What materials, type of toys, and composition of the toys themselves have been changing since the last three decades? Related reading: *TIME, Discover.*

52. How have air quality regulations affected small businesses in the United States? Related reading: *TIME.*

53. How has weather affected transportation and disposal of hazardous waste material by the trucking industry? Related reading: *TIME, Discover*, or contact the Department of Transportation in Washington, D.C.
54. How does weather affect the quality and the accessibility of building materials for new home construction? Related reading: *National Geographic*.
55. Suppose a giant black-colored plastic shield (200 miles by 300 miles by 2 feet thick) were sunk off the coast of Southern California in 100 feet of sea water. Could this result in evaporation of water that could assist in keeping inversion layers away from the San Fernando Valley or L.A. Basin by generating more summer thunderstorms? How could this help in pollution reduction? What are some potential extremely detrimental effects? Is there an experiment that could be designed to model at least a part of this idea? Related lab: 16.
56. How does weather affect the removal, transportation, and subsequent dumping of waste in landfills? Related reading: *National Geographic, TIME*.
57. Is there any link to "ionized air" and human/animal behavior? Related reading: *Psychology Today, OMNI*.
58. So-called room ionizers remove pollutants. How do these machines function? Are they effective in clearing a household's breathing atmosphere? Related reading: *Discover*.
59. In the sport of hot-air ballooning, a pilot/navigator must "see ahead" certain geophysical features in order to safely operate the balloon. How would a balloonist navigate safely at a low altitude (less than 1000 feet) over a large grassy or concrete area? Related labs: 3, 4, 6, 13, 16, 19, 20, 21.
60. How have enclosed playing areas, "domes," changed the playing of sports in areas of poor weather? What are "ideal" conditions (microweather) inside of these domes? How does the financial outlay compare to the advertisers supporting a game that isn't "called on account of rain?" Related labs: 2, 3, 4, 5, 6, 7, 17.

Appendix D
Required equipment

Equipment required for all laboratory exercises

meter stick
straight pin
white typing paper
glue stick
hand-held hole punch
scientific calculator
cardboard
50-meter measuring tape
large square mirror (6-inch by 6-inch)
scissors
colored pencils
30 Celsius thermometers
microwave oven
hot plate
hot air popcorn popper
loose popcorn
popcorn in microwave bag
popcorn in stove top container
lamp with 60, 75, or 100-watt bulb
masking tape
graph paper
watch or clock with second hand
5 small plastic bowls
5 plastic centimeter rulers
glass aquarium
plastic wrap
glass cover (for aquarium)
beakers: 500, 250, 150, 50 ml

nylon hose
flour
cobalt chloride crystals
table salt
distilled water
untreated white cotton cloth
black poster board
electric fan
2-liter plastic soft drink container
candle
newspaper
matches
smoke chamber
6 sand bags (to hold 500 ml of sand)
ringstand
test tube clamp
10 large test tubes
10 one-hole rubber stoppers
untreated steel wool
overhead projector marker
drinking glass
8 plastic cups of different colors
white paper towels
samples: salt water, dirt, sand, grass
small wood blocks
calcium chloride (powdered)
VHF weather radio
plasticized white board

ice cubes
toothpicks
3-by-5 cards
white glue
rubber bands
wide-mouth glass jars
uninflated balloons

½-inch self-adhesive letters
T square
paper cutter
3-foot-by-5-foot corkboard
erasable markers (red, green, blue)
goggles
heavy work gloves

Appendix **E**

Manufacturers of laboratory equipment

In addition to writing or calling the following companies for catalogs or other information, you can also check with your local high school or college. (Call a department chairperson to obtain information on borrowing equipment or chemicals for your own school.)

A & A ENGINEERING
2521 W. La Palma Ave.
Anaheim, CA 92801
(714) 952-2114
Electronic equipment: power supplies, satellite receivers, decoders.

CAROLINA BIOLOGICAL SUPPLY CO.
2700 York Rd.
Burlington, NC 27215
(800) 547-1733

EDMUND SCIENTIFIC CO.
101E. Glouchester Pike
Barrington, NJ 08007-1380
Ordering Number:
(609) 573-6250
Product Information:
(609) 573-6259

FISHER SCIENTIFIC
4901 W. LeMoyne St.
Chicago, IL 60651
(800) 621-4769

FREY SCIENTIFIC CO.
905 Hickory La.
Mansfield, OH 44905
(800) 225-FREY

RADIO SHACK
—for inexpensive VHF weather radios
(Check local listings in white pages of your directory or call information.)

SARGENT-WELCH SCIENTIFIC CO.
7300 North Linder Ave.
PO Box 1026
Skokie, IL 60077
(312) 677-0600

SCIENCE KIT & BOREAL LABORATORIES
777 East Park Dr.
Tonawanda, NY 14150-6782
(800) 828-7777

Appendix F
Answers to lab questions

Because regional weather and the equipment used to simulate it are quite variable, the following answers to the questions at the end of each lab will be somewhat different for your own classroom and region.

All measurements, data for graphs, and subsequent plotting of data occurred at Westlake High School in Thousand Oaks, California. The labs were performed at room temperature (21 degrees Celsius), or outside on a sunny day (unless the experiment procedure called for a different situation).

Use these answers as a guide and keep in mind that your data and analysis of it are just as valid as the answers presented here.

LAB 1: DIAMETER OF THE SUN
Questions/conclusions
1. This answer will depend on your individual measurements and calculations.
2. Gravitational and rotational (centripetal) forces from the center of a planet or star pull the mass of that body equally toward its center. This produces a spherical-shaped object. On another note, it's the spin on the earth that gives the earth a slightly squashed shape.

LAB 2: ISOTHERMS IN THE CLASSROOM
Observations
1. Temperature increases as elevation increases because the warm air from any heating units will travel to the highest point in the room. The warm air is also less dense than the cooler air near the floor and desks.
2. The temperature will vary for a number of reasons: outside weather, inside air

currents from any open door or window, location of heating/cooling ducts, time of day, and number of students in the room.

Questions/conclusions

1. It's similar in the use of smooth lines and plotted equal temperatures. It's different because the U.S. map is for a larger region, the method of gathering the temperatures for the U.S. map is different than the students', and the elevations used by weather balloons are different.
2. Weather balloons, National Weather Service observation stations, sea buoys, and amateur weather observers.
3. To make the air inside the balloon less dense than the cooler, denser air outside the balloon. When this occurs, the balloon will begin to pull against gravity and rise.

LAB 3: RADIATION, CONVECTION, CONDUCTION
Observations

1. The bag used in the microwave is the coolest. Radiation from the microwave heats only the food and not the container that it occupies. Microwave radiation was able to penetrate this paper bag; the bag was "transparent" to the radiation.
2. The container used for the conduction portion was the hottest (immediately after the popcorn popped). The container was in direct contact with its source of energy; the heat first penetrated the container to reach the popcorn. The heating element, the container, and the food all transferred their heat/energy to each other through vibrating molecules.

Questions/conclusions

1. Through radiation from the sun.
2. By converting the shortwave radiation from the sun into longwave radiation (inside the greenhouse). This longwave radiation can't escape the structure and heats the interior.
3. A mirage (an optical illusion) is formed when light rays are refracted (bent) as they strike a boundary between layers of hot and cool air. A superior mirage is an image that seems to be suspended in the sky. An inferior mirage is an image that appears to be on the ground. The layer of cool air reflects the sky onto the ground.
4. Heating by convection.
5. A fireplace exhibits all three forms of energy: radiation warms a person from a distance, hot air (convection) currents travel up the chimney, and a metal poker becomes quite warm when "stoking" the fire.
6. A hot stove top is almost instantly "hot" to the touch. Human reflexes quickly pull a hand or arm away from the danger. Microwave radiation damage is nearly painless, and by the time pain is felt, the damage from the microwaves can be quite severe. Although microwave ovens are extremely safe when used

properly, transmitters and their antennas (for mobile phone communications as well as amateur radios) can give off hazardous microwave energy.

LAB 4: HEAT IN THE ATMOSPHERE
Observations

1. The thermometer closest to the light source.

Questions/conclusions

1. The Northern Hemisphere is tilted away from the sun; the rays of light from the sun are not striking the earth as directly as they would in the winter.
2. The 23½° tilt of the earth causes it to receive different amounts of sunlight (and heating) throughout the year.

LAB 5: TEMPERATURE & EVAPORATION OF WATER
Observations

1. The amount would be proportionally the same if all other factors remained constant.
2. Yes, the amount of exposed surface area to the heat source would be different.

Questions/conclusions

1. Heated air over the desert floor evaporates the falling rain.
2. The equatorial regions. There's more direct sunlight received in these regions throughout the year.
3. It's a percentage of solar radiation reflected by a surface.
4. The darker a surface is (the higher its albedo), the hotter the surface will become with sunlight striking it. The heating of this surface will cause any moisture to evaporate more quickly than a lighter colored surface. Lighter colored surfaces (surfaces with a lower albedo) reflect more light than darker colored surfaces.

LAB 6: HEATING BY CONVECTION
Observations

1. Answers will vary depending on the size of the tank and the time the air in the tank is heated/cooled.
2. After ten minutes, thermometer number nine is the coolest. (See graph.) The warmest region is by thermometer number one.
3. According to the given data, no. Answers might vary depending on the length of time the experiment is run.

Questions/conclusions

1. Movement of air caused by uneven heating/cooling. Warm air rises and is replaced by cooler, denser air.

2. By bringing parcels of heated/cooled air over a given region.

3. It's local weather. Air fans used in a fruit orchard on a cold night will prevent the freezing of a crop. Gloves worn by skiers to keep the hands warm are a form of microweather. The predictability of microweather is determined by the size of the region involved, whether it's a pocket in a jacket, an orchard, or a city.

LAB 7: DETERMINING DEW POINT
Observations

1. It takes time for the air to become saturated. When condensation formed, the air inside the beaker and above the ice was saturated with water vapor.

2. Yes, after the glass exterior "fogged up."

Questions/conclusions

1. The height (in meters/kilometers or feet) where an air parcel forms into a visible cloud.

2. The lifting condensation level (LCL) is the point where the parcel of air is saturated with water vapor and is cool enough to become visible.

3. Ground fog/radiation fog: forms when the nighttime sky is clear and the ground cools rapidly due to loss of heat by radiation. When winds mix the cold bottom air with air just above the ground, this layer is cooled below the dew point, and fog forms. Small air currents keep these tiny fog droplets aloft. Such fogs form commonly near rivers and lakes in humid valleys. They occur mostly in the fall period. They're thickest in the early morning and dissipate by the morning rays of the sun.

4. The relative humidity in a ground/radiation fog is at or near 100 percent.

LAB 8: THE BAROMETER
Observations

1. The weather was mostly unsettled; cloudy with drizzle/rain or fog.

Questions/conclusions

1. Aneroid barometer: an instrument that measures air pressure with a partial vacuum in a sealed metal chamber. The sides of this chamber bend inward or outward depending on the pressure changes. Aneroid means "without liquid."

2. Mercurial barometer: pressure from the atmosphere presses on liquid mercury (element symbol Hg) in a well at the barometer's base. The height of the mercury column depends on the atmospheric pressure. Six different ways of expressing standard pressure at sea level are: 760 millimeters Hg, 1014 millibars, 101.4 kilopascals, 1 atmosphere, 14.7 pounds per square inch, and 29.92 inches Hg. The mercurial thermometer tends to be somewhat more accurate in indicating atmospheric pressure changes since any temperature changes surrounding the metal chamber of an aneroid barometer can affect its reading.

LAB 9: CALCULATING THE SIZE OF RAINDROP
Observations
1. Depending on the wind direction and how the container was held, the shapes varied from nearly circular to elliptical for the larger drops.

Questions/conclusions
1. Condensation nuclei: (ice or dust particles), less than 0.001 millimeter in diameter. Fog: suspended in air due to light winds (approximately 0.02 millimeter in diameter). Drizzle: form of liquid precipitation, less than 0.5 millimeter in diameter, but larger than 0.02 millimeter. Rain: 0.5 to 5 millimeter in diameter, drops larger than 5 millimeters break up as they fall to the ground.
2. Larger drops fall faster than smaller drops; there's then a tendency for smaller droplets to get "sucked in" behind the bigger drop and collide with it to increase the size of the larger drop. Still, larger drops can collide with smaller drops as they fall, capturing them. Sometimes the smaller droplets bounce off the larger drops without any capture.
3. Collision: see answer number two. Collision is known as a warm-cloud process. Ice crystal: except for low-lying clouds in the tropics, upper layer cloud temperatures are below the freezing point of water (0° C). Supercooled water in these clouds is deposited on the ice crystals. When heavy enough, the ice crystal falls and picks up more water droplets on the way toward the surface. If the temperature in the lower portion of the cloud is above freezing, the crystal melts and continues to fall, growing in accordance with the collision process.
4. They all need a piece of dust, smog, salt or chemical (silver iodide) to begin the formation of a droplet.
5. It provides condensation nuclei, which are tiny bits of unburned particles of smoke for water molecules to stick to.

LAB 10: THE RELATIVE HUMIDITY INDICATOR
Observations
1. It was a pink color.
2. Approximately twelve hours to two days.

Questions/conclusions
1. One simple device is a hair hygrometer. It measures relative humidity of the air by using a human hair with a pointer attached to one end. The other end is fixed. When moisture is present (high humidity) the hair stretches. When the air is dry, the hair shrinks. In both cases, the indicator responds accordingly.
2. This instrument also measures the relative humidity of the surrounding air. It's based on the notion that evaporation causes cooling. Two thermometers are mounted side by side on a plastic or wooden panel. One of the bulbs is wrapped in a water-soaked cotton wick. The thermometers are either whirled or

fanned so air is moved by each of the bulbs. As the water evaporates from the cotton wick surrounding the bulb, heat is taken away. In very dry air, the evaporation happens quite quickly, indicating a low temperature on the wick-covered bulb. On rainy or very humid days, the "wet bulb" thermometer changes little. These two thermometers don't indicate the relative humidity, however. They only show how dry the air is surrounding them. With these two readings, the relative humidity can be determined by using a table.

3. On a rainy day, the relative humidity of the air is at or near 100 percent. A sweaty person would not cool off as quickly—the sweat on the skin would not evaporate nearly as fast as it would if the person were in a room on a hot, dry summer day.

LAB 11: THE WIND CHILL FACTOR
Observations

1. Yes, over the indicated time period.
2. Yes, at minute three and at minute eight.
3. For this data and subsequent graph, there was no steady plateau.

Questions/conclusions

1. Depending on how hot the temperature of the water initially was, the thermometer would indicate just the cooling of the water. After reaching room temperature, the evaporation/wind chill curve would be observed.
2. As you climb out of the pool, dry air and light winds "strip away" water molecules from your skin. Each of these molecules also takes with them a "piece" of heat that your body generated. Evaporation is a cooling process and your brain registers your skin being colder.

LAB 12: THE WEIGHT OF THE ATMOSPHERE
Observations

1. For several minutes after the cap was twisted on, the container crumpled and collapsed.
2. By blowing into it (with lung power) or by filling it (slowly) with tap water.
3. A great deal less than the outside air pressure.

LAB 13: A CONVECTION CYCLE
Observations

1. It travels into the opening that's without the lit candle, back out over the candle's position, and out of the chamber.
2. It is to create a region of different temperatures, so that a "wind" can travel.
3. The high pressure is at the position where the smoke enters. The low pressure is over the candle where the smoke exits the chamber.

Questions/conclusions

1. Hadley Cell: a wind that replaces the rising air over the doldrums (the inter tropical convergence zone—ITCZ.) The Hadley Cell completes the convection system at the equator and was named after George Hadley, its discoverer in the 18th century. Your experimental "cell" also demonstrates air rising by convection.
2. By uneven heating/cooling on the earth's surface.
3. (See the previous answer.)
4. Winds flow from high pressure areas to low pressure areas.
5. The kite will fly away from the water and shore at noon. It will fly over the water away from the sand in the early evening.

LAB 14: THE COLD FRONT
Observations

1. Depending on the size of the tank, condensation formed above the sand bags to near the top of the aquarium.
2. Just below the condensation line.

Questions/conclusions

1. It's a leading edge of a mass of cold air that strikes a parcel of warm air. The advancing front has a very steep frontal surface.
2. The frontal surface of a cold front is steep near the surface, primarily because friction at the ground slows down the movement of the cold air. Warm air is pushed skyward where it cools. Eventually, the lifting condensation level (LCL) of this warm air parcel is reached and condenses into large cumulus and cumulonimbus clouds. Storms created along a fast-moving cold front are quite violent; a squall line—a band of heavy thunderstorms advances just ahead of a cold front. Slow-moving cold fronts produce less concentrated cloudiness and less precipitation.

LAB 15: THE OCCLUDED FRONT
Observations

1. Depending on the location of the bags and the size of the aquarium, condensation formed over the "sandwich" of sand bags and high over the hot water.

Questions/conclusions

1. An occluded front forms when a front of cold air overtakes a front of warm air and lifts this parcel of warm air completely off the ground. The warm air is then cut off or occluded from the surface. Cool air that develops beneath the lifted warm air parcel comes into contact with the advancing cold front. The warm air is parallel to the earth's surface.

2. Precipitation from middle to lower-level clouds might result because of an occluded front.

LAB 16: THE TEMPERATURE INVERSION

Observations

1. The layer of smoke formed over the sand bags and the beaker of smoldering newspaper, but slightly above the beaker. If closely watched, the smoke bands formed into thin lines composed of alternating air and smoke.
2. The heat from the lamp pushed down on the band of smoke.
3. The cooler temperatures from the sand bags with the heat from the lamp trapped the smoke in the aquarium creating this limited "inversion."
4. To simulate the cooler moist air that's needed to produce a temperature inversion layer.

Questions/conclusions

1. Cool, moist air from the Pacific moves into valley regions, becomes polluted, and is then covered by a warm air layer. Mountains that border the valley prevent any circulation by the air. This layer traps additional automobile exhausts and becomes smog (smoke and fog).
2. Strong winds can prevent the formation of a temperature inversion. Or, if the sky remains clear, the early morning rays of the sun warm the ground and the lower atmosphere. The low-level inversion is partially cleared by late morning.
3. Other cities with basins, and beside mountain ranges.

LAB 17: MEASURING OXYGEN CONTENT OF AIR

1. Balanced equation: $4Fe + 3O_2 \longrightarrow 2Fe_2O_3$
2. The rusting of metal objects can be limited by using protective paints, enamel coatings, or plastic coverings.

Observations

1. Prior to the experiment, the steel wool should have a gray color. After the experiment, a dark orange to brown color is seen in patches throughout the wool. This was a chemical change.
2. The wool was dampened with water to allow the rusting of the iron to occur more quickly inside the test tube.

Questions/conclusions

1. Answers will vary depending on how carefully the student performed the experiment and how accurately the markings were measured on the test tube.
2. The water level will reach its highest point after about three or four days (depending on the wool being dampened prior to the experiment). In any size

sample of air, there's a fixed percentage of oxygen that will completely chemically combine with the iron in the steel wool. In chemical terms, the iron is oxidized and the oxygen is reduced.

LAB 18: AIR EXERTS PRESSURE

Observations

1. Outside air pressure far exceeds the interior water pressure of the glass.
2. Air pressure is pressing on all sides of the glass, and is much stronger than the weak water pressure (which is also pressing on all sides of the glass).

Questions/conclusions

1. Air pressure: the weight of the atmosphere per unit area. At sea level, standard air pressure is expressed in a variety of ways. 760 millimeters mercury (Hg), 29.92 inches mercury, and 1014 millibars are the most common. Air pressure is due to the sum total effect of molecules of air colliding against any surface.
2. As altitude increases, there are less air molecules colliding (with the instrument measuring such high altitude pressure), and the pressure is lower.
3. If the experiment were performed in an airborne pressurized jet cabin, most likely the square would stay on just as easily as it did on the surface. On a mountain top, the square would not remain on the glass as well as it did on the surface. There would be less air molecules striking the square.

LAB 19: HEATING AND COOLING OF MODEL LAND FORMS

Observations

1. Depending on the colored cups used by the student, the darker-colored cups should heat the fastest and the lighter-colored cups should heat the slowest.
2. Depending on the colored cups used, this answer will vary. For the data used in this example, there were plateaus for the black, dark green, and red. The white and light green cups also had a notable plateau.
3. The paper towels prevented direct (sun)light from striking the thermometer bulb. The heating was mostly due to the cup's absorption of the radiation from the light source. If the towels had not been in the cup, the heating would have occurred much faster.

Questions/conclusions

1. Sand, desert, dirt, rock, and grassland are five possible land forms.

LAB 20: HEATING AND COOLING OF ACTUAL LAND FORMS

Observations

1. From the data obtained for this lab, the sand heated up the fastest. The fresh water heated up the slowest.
2. The grass blades cooled off the fastest (there's a steeper slope from minute 10 to minute 15). The sand cooled off the slowest.

3. The salt water heated up slightly faster than the fresh water. They both had similar cooling slopes.

Questions/conclusions

1. Sand is more densely packed than water molecules. Heat is transferred more quickly from each sand grain with sunlight striking them. Water heats up much more slowly because it has a lower specific heat (it can absorb a great deal more heat energy than sand without an appreciable temperature change).

LAB 21: HOW MOUNTAIN RANGES AFFECT CLIMATE
Observations

1. Depending on the size of the aquarium, the side with the beaker showed signs of dampness first.
2. Over a period of several minutes, the mountain became saturated with "rain."
3. The windward side had the most moisture deposited on it.

Questions/conclusions

1. Orographic uplift—a wind that's mountain-modified. Moist parcels of air that are pushed up the windward side of a mountain condense into clouds and then dump their moisture in either snow or rain. The leeward side tends to be drier than the windward side.
2. Lee side—the side of a mountain range facing away from orographic winds. Barren landscape leading to a desert region is usually found here.
3. Windward side—the side of a mountain range that faces the orographic winds. The landscape is quite lush with foliage.

LAB 22: LONG-TERM WEATHER OBSERVATIONS AND GRAPH
Observations

1. Any observed trends are dependent on the student's data. For the data here in this example, the temperature was gradually decreasing.
2. This answer depends on the observed data. This example graph was for the middle of winter.
3. Not necessarily. Weather in the Los Angeles area is quite variable, and the winters can be quite mild.
4. This answer depends on the observed data. Here, the humidity readings fluctuated a great deal.

Questions/conclusions

1. For the data obtained here, it was possible to forecast an unsettled period and not necessarily precipitation. (Southern California weather is the most difficult to predict accurately in the United States.) Depending on a student's data, yes, there should be precipitation following within one to two days.

2. Precipitation will generally occur with a drop in barometric pressure, a rise in humidity, and a shift of wind direction to an easterly course. Fair weather will generally occur with a rise in barometric pressure, a drop in humidity, and a shift of wind direction from an easterly to a westerly course.

LAB 23: SATELLITE PICTURES FROM THE NEWSPAPER
Observations

1. The answer will depend on the location and season.
2. In the Northern Hemisphere, they travel from west to east.

Questions/conclusions

1. Yes, the prediction will be made much more accurately if regional readings of temperature, pressure, humidity, and cloud type are used with the satellite photographs.
2. This answer depends on latitude. See a diagram of prevailing winds for the globe in any earth science text.
3. The velocity (speed and direction) of the front will be given over television, radio or newspaper. Obtain a map of the United States (available at local automobile travel clubs). Locate the center of the front on the map. For the plotting, be sure to use the map's legend indicating distances. The front's next location can be approximately predicted.

   For example: A front is moving at 12 mph (19.3 kilometers/hour) in an easterly direction. In 24 hours, the front's location will be 288 miles (463.5 kilometers) from its original location. If city B is 524 miles (843.3 kilometers) east of city A, the front will reach city B in almost two days. Use the formula:

$$\text{distance (miles or kilometer)} = \text{rate (mph or km/hr)} \times \text{time (hrs)}$$
$$288 \text{ miles} = 12 \text{ mph} \times 24 \text{ hours}$$
$$\text{time (hours)} = \frac{\text{distance}}{\text{rate}}$$

$$44 \text{ hours (1.8 days)} = \frac{524 \text{ miles}}{12 \text{ mph}}$$

LAB 24: "SPOT" 24-HOUR WEATHER PREDICTIONS
Observations

1. The answer will vary for different regions, but there should be some accuracy in each forecast (a typical accuracy is about 70% to 75% correct).

Questions/conclusions

1. The information obtained for NWS (National Weather Service) forecasts is obtained from balloons, observers, remotely located sea buoys, satellites, and radar. The information typically includes: wind speed at various heights, humidity levels, dew point, water temperatures, and locations of other influencing fronts.

2. Computers have quickened the compilation of data and production of weather maps. Satellites maintain a constant vigil watching over the earth's cloud patterns and movements. Both instruments have reduced the tedium of calculating and making "blind" forecasts. The meteorologist has at his or her disposal very powerful tools to make highly accurate forecasts.

Index

A
absorption of heat, 78-88
air-quality (pollution) index, 110
answers to lab questions, 132-143
atmosphere and weather, 37-76
  barometer to make and use, 38-41
  cold fronts, 61-64
  convection cycles in heated air, 27-30, 58-60
  graphing wind direction and related factors, 115-116
  occluded fronts, 64-67
  oxygen content of air, 71-73
  pressure of air, 74-76
    graphing trends, 114-116
  raindrop sizes, 42-45
  relative humidity
    indicator, 46-49
    tables/graphs, 117-118
  squall lines, 61
  temperature inversions, 68-70
  weight of atmosphere, 56-57
  wind chill factors, 50-55
  wind direction and speed graphing, 118-119

B
barometer to make and use, 38-41
barometric pressure, 74-76

C
charts (see graphing weather trends)
cold fronts, 61-64
conduction of heat, 12-17
convection of heat, 12-17, 27-30, 58-60

D
dew point, 31-35

E
earth's effects on weather (see also geophysical features), 23-35
  convection cycles in heated air, 27-30, 58-60
  dew points, 31-35
  heating and cooling of land forms, 78-88
  mountain ranges and climate, 89-91
  water: temperature and evaporation, 24-26
environmental question-and-answer lab, 123-127
equipment required for labs, 128-129
  manufacturers and suppliers, 130-131
evaporation of water, 24-26

F
forecasting weather, 93-106
  hurricane tracking, 108
  long-term weather observations and graph, 94-98
  pollution and air-quality index, 110
  rainy seasons, 110
  satellite pictures, 98-101
  snowstorm tracking, 110
  "spot" 24-hour predictions, 102-104
  thunderstorm watch, 109-110
  tornado tracking, 108-109
  tracking seasonal storms, 107-110
  trends in weather, (see also graphing weather trends), 111-122
  weather-data stations, 105-106
fronts
  cold fronts, 61-64
  occluded fronts, 64-67

G
geophysical features (see also earth's effects on weather), 77-91
  heating and cooling of land forms, 78-88
  mountain ranges and climate, 89-91
graphing weather trends, xiv-xv, 111-122
  air pressure and related factors, 114-116
  long-term weather graph, 94-98
  precipitation amounts, 120-122
  relative humidity and related factors, 117-118
  temperature table, 112-113
  weather statistics charts, 119
  wind direction and related factors, 115-116
  wind direction and speed, 118-119
groups, lab groups

equipment required, 128-129
  manufacturers/suppliers, 130-131
length of time for labs, xv
size of group, xv

H

humidity, relative humidity, 46-49
  graphing trends, 117-118
hurricanes, tracking hurricanes, 108

I

inversions, temperature inversions, 68-70
isotherm mapping, 7-11

L

length of time for labs, xv

M

moon, diameter of the Moon, 5-6
mountain ranges and climate, 89-91

O

occluded fronts, 64-67
oxygen content of air, 71-73

P

pollution index for air quality, 110
popcorn, popping corn with different heat sources, 17
precipitation, graphing amounts, 120-122
pressure of air, 74-76
  graphing air pressure and related factors, 114-116

R

radiation of heat, 12-17
rain (*see* precipitation)
rainy-season tracking, 110
reflection of heat, 78-88
relative humidity, 46-49
  graphing trends, 117-118

S

safety procedures, xiii-xiv
satellite weather observation, 98-101
size of lab groups, xv
snow (*see* precipitation)
snowstorm tracking, 110
"spot" weather forecasting, 102-104
squall lines, 61
storm tracking, 107-110
  hurricanes, 108
  pollution and air-quality index, 110
  rainy seasons, 110
  snowstorms, 110
  thunderstorm watch, 109-110
  tornadoes, 108-109
Sun and solar experiments, 1-22
  diameter of the sun, 2-6
  isotherms and temperature layers, 7-11
  radiation, convection, conduction of heat, 12-17
  temperature vs. distance from heat source, 18-22

T

temperature-related labs
  absorption of heat, 78-88
  conduction of heat, 12-17
  convection of heat, 12-17, 27-30, 58-60
  dew point, 31-35
  evaporation and heat, 24-26
  graphing temperature trends, 112-113
  heating and cooling of land forms, 78-88
  inversions, 68-70
  isotherm charting, 7-11
  popping corn with different heat sources, 17
  radiation of heat, 12-17
  reflection of heat, 78-88
  temperature vs. distance from heat source, 18-22
  wind chill factors, 50-55
thunderstorm watch, 109-110
tornadoes, tracking tornadoes, 108-109
tracking seasonal storms, 107-110

W

water-related labs
  dew point, 31-35
  evaporation and heat, 24-26
  precipitation, graphing amounts, 120-122
  raindrop sizes, 42-45
  relative humidity indicator, 46-49, 117-118
weather forecasting (*see* forecasting weather)
weather statistics charts, 119
weather-data station, 105-106
wind
  chill factors, 50-55
  direction and speed graphs, 118-119